ELECTROMAGNETIC SPECTRUM

NEW

MANY SPEEDS THEORY

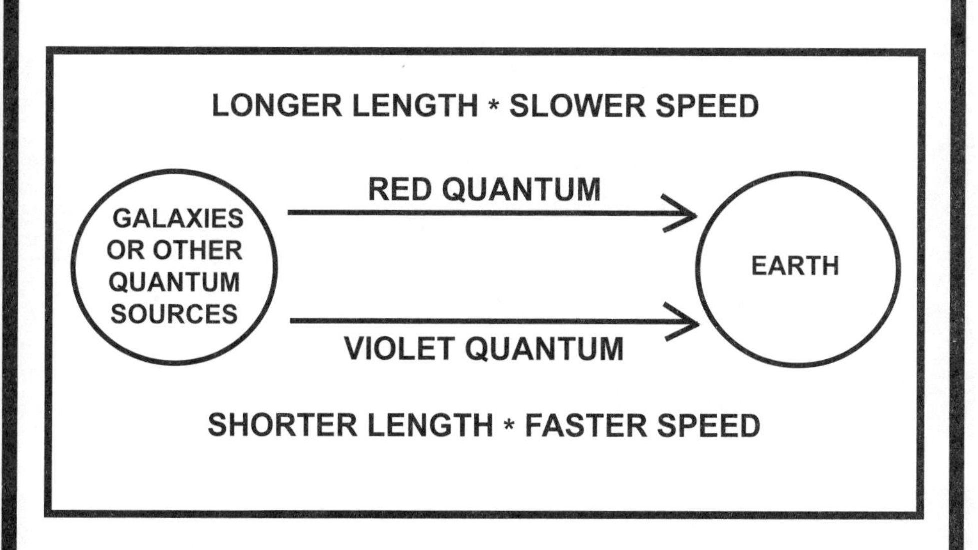

Walter H. Volkmann

Every effort has been made to avoid errors, but despite best efforts, some will appear. Please bring these errors to my attention, so they can be corrected in future editions.

First Printing 2008

Volkmann, Walter H.

ELECTROMAGNETIC SPECTRUM – NEW –
MANY SPEEDS THEORY

Includes Index

ISBN 978-0-9765228-1-2

Printed in the United States of America
by Taylor Publishing Co. Dallas, Texas 75235

Table of Contents

INTRODUCTION TO THE BOOK

This book is for thinkers, those who have an inquisitive mind. Think each sentence as you read, many times one sentence will have lot of information. The book will not have a lot of superfluous writing or information.

In this book, you will find mostly new theories that are not to be found in the textbooks, and as far as I know, are not to be found anywhere. It may be difficult to understand and accept the new theories because they are new and unfamiliar.

I have included many line drawings for better understanding of the written word. There are no hyphenated words at the end of a line to interrupt one's thoughts with trying to decide, "what is that word?"

This book has been a very laborious undertaking. A 100 times more than I thought at the start. It took almost all my extra hours. I was always short of hours to write and rewrite.

Many nights I would get up at 1:30 a.m. to study, going back to bed at 3:30 a.m. Everything was quiet, my mind was at it's best. Ideas seem to develop, racing through my mind as I was reading. Many good ideas were lost, due to the fact could not write them on paper fast enough. Before one thought was written, others were going through my mind.

It was an experience that happened many times. After

the ideas were developed, I could hardly believe that this happened. The events that occurred were mysterious. I cannot fully explain.

Flower Mound, Texas W.H.V. 2008

Acknowledgements
To

Jay Love - Taylor Publishing Company - Dallas, Texas
Thom Cashman - Dallas Design & Printing - Dallas, Texas
Johnny Gomez - Graphics by Gomez - Dallas, Texas
Jennifer Hannah - Graphic Designer - Carrollton, Texas
Scott Young - Scott Young Photography - Flower Mound, Texas

Many thanks to the above named persons who helped with their professional knowledge & expertise to produce this book.

Mr. Love explained the book publishing process and sent me to Mr. Cashman to transform my hand-printed notes into a form that could be used to publish a book. 1945 - 2008

Mr. Cashman is a person with a lot of knowledge on preparing an author's work into a manuscript for the publisher to print a book. This requires a lot of patience trying to please the author and the publisher - a talent that he has.

Mr. Gomez is a professional graphic artist, who took my sketches and made them into the fine drawings you see in the book.

Mrs. Hannah is a graphic designer, that works at Dallas Design and Printing. She created Drawings 2-1, 2-2, 4-1, 4-2, 5-1, 11-1, 11-2, 11-3 and 14-2 from my sketches.

Mr. Young came to my home to take pictures of me in my study, so you the reader, could see the author and where he wrote his book.

And last of all, my wife Jeanette for her patience and help with typing certain parts of the book.

About the author

BIOGRAPHY

Walter H. Volkmann
Scientist

I have been in many fields of science since my early school years. In high school, I excelled in chemistry and physics. When I graduated from high school, my physics teacher tried to get me a scholarship, but due to the war, nothing could be done at the time.

My first job was at a testing laboratory, which tested cottonseed and its products. My second job was at a flour mill as a quality control chemist.

Then I had the opportunity to go to college in San Francisco, California. While going to college, I worked at a laboratory doing tests on alfalfa. Here I tested for carotene, using chromatography columns.

After college, I came back to Dallas, Texas. Being interested in radio, I studied for my amateur radio license, passed the test and received my license with the call letters W50MJ. I then studied for a commercial radio license and passed the test. With the extra knowledge of the commercial radio license, I was qualified to work on aircraft electronic equipment. I got a job with one of the large airlines.

Then I had the opportunity to go to work at a blood research center and as a chemist in the special chemistry laboratory at the associated hospital.

I was called into the Army, and was sent to their school for special training to help run one of their medical laboratories.

Afterwards I decided to go into business with my brother in the plant growing business. This business specialized in Saint Paulia (African Violets). At the time, growing plants by the tissue culture method was starting. To learn this method, I went to school for training at the Jones Cell Science Center at Lake Placid, New York. I went on to study at Pennsylvania State University to learn "Plant Biotechnology Methods", and following that, I went to the Center for Advanced Training in Cell and Molecular Biology.

I set up a plant tissue laboratory to grow plants by the tissue culture method. Before this, plant tissue culture training. I took a course called "McGraw-Hill Contempory Electronics".

Now, here I am 84 years old, and writing this book!!

Walter H. Volkmann

NOTES

CHAPTER 1

INTRODUCTORY INFORMATION

Presenting an introduction to a new world. The electromagnetic spectrum quantum forming. The quantum is a packet of electromagnetic energy (light, radio, x-rays, etc.) Formed by inter-atomic changes within the atom.

A new many speeds theory on the quantum formed by inter-atomic atom changes and the amount of energy of each quantum.

In my theory I have made some assumptions as all scientists have done in presenting new theories.

According to Max Planck's quantum theory each quantum has a different energy value. In my theory each quantum has the same energy value.

According to Albert Einstein, in his theory of special relative, all electromagnetic radiation (quanta) travel at the same speed. In my theory each quantum travels at a different speed.

ELECTROMAGNETIC SPECTRUM – NEW – MANY SPEEDS THEORY

The prevaling theory in science today (2008) is that the speed of all parts of the electromagnetic spectrum - visual (light), radio, x-rays, etc. have the same speed in a vacuum (space).

You have been taught this "cradle to grave"; so it must be true – right – well mabye not. Read my theory on the subject and decide for yourself.

Studies begain to put doubt in my mind on the validity of this one speed theory. In my theory the electromagnetic spectrum is not all one speed, but many different speeds in a vacuum (space).

The English scientist Issac Newton (1642-1727) thought light was a stream of particles. The English scientist Thomas Young (1773-1829) demonstrated with his double-slit experiment that light had the characteristics of a wave, so; now we have light both particle and wave. The Danish astronomer Olaf Roemer (1644-1710) determinded the speed of light. His method found how long it took light to travel the major axis of the Earth's orbit. Another English scientist James Clerk Maxwell (1831-1879) believed in the wave theory and came up with light waves being part of the electromagnetic spectrum. Some experiments show light to be an electromagnetic spectrum wave and other experiments show light to be a particle.

The German scientist Max Planck (1858-1947) discovered by his experiments that light was in the form of packets of energy, called quanta and so from this the Quantum Theory was born, for which he received the Nobel Prize. In 1905, the

German Scientist, Albert Einstein (1879-1955) explained the photoelectric effect confirming the quanta concept, for which he received the Nobel Prize.

I am not a believer in the wave theory for the electromagnetic spectrum in its normal state. However, there are man-made electromagnetic radiation, radio being one. This man-made radio radiation if put on a wire whose length is in resosance with the radiation – and – then electrical measurements of voltage and current can be made on the wire – and – when these measurements and are plotted on an x-y graph the curves will be a sine wave.

In this book I will use th word quantum length instead of wave length and quantum instead of photon. Quanta means packets of energy. A term from quantum theory – that is what we have – "quanta" – packets of electromagnetic energy from within the atom. No mass.

Now 350 years later it appears that Newton's Stream of Particles was a move in the right direction, only now it is Quanta "Packets of Electromagnetic Energy".

In the 1880's Maxwell with his wave theory and mathematics was accepted by science. Then in 1900 Planck came forth with his Quantum Theory. Light was packets of energy called quanta.

What happened to the Wave Theory? If light is a wave then what is making light and other electromagnetic radiation – in its normal state – a wave? The wave theory is based mainly on Young's double slit experiments and Maxwell's theory.

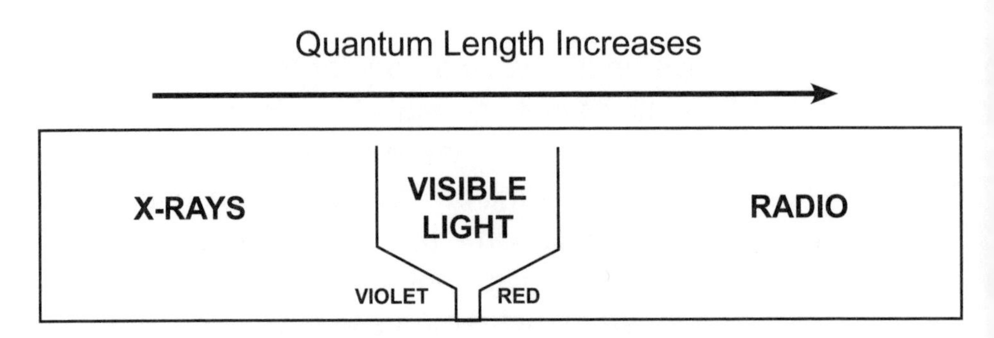

Electromagnetic Spectrum

CHAPTER 2

ELECTROMAGNETIC PHYSICS

Each element produces its own specific spectrum of quanta. With this information a light spectrum can be analyzed to determine what elements made the spectrum. This great analytical tool was discovered in the 1860's by the German scientists Gustav Kirchoff and Robert Bunsen.

Light spectrograms from our sun, stars, galaxies and other light sources can be analyzed to determine what elements make-up the light source.
FOR TEXT REFERENCE PLEASE REFER TO DRAWING 2-1. Here we can see that the quanta lengths are different for each quantum. The quantum lengths become shorter as the inter-atomic quanta forming times become shorter.

Each quantum length has the same energy value. The red quantum length has all its inter-atomic forming energy value in a longer quantum length, (compared to the violet quantum length) while the shorter violet quantum length has the same energy value.

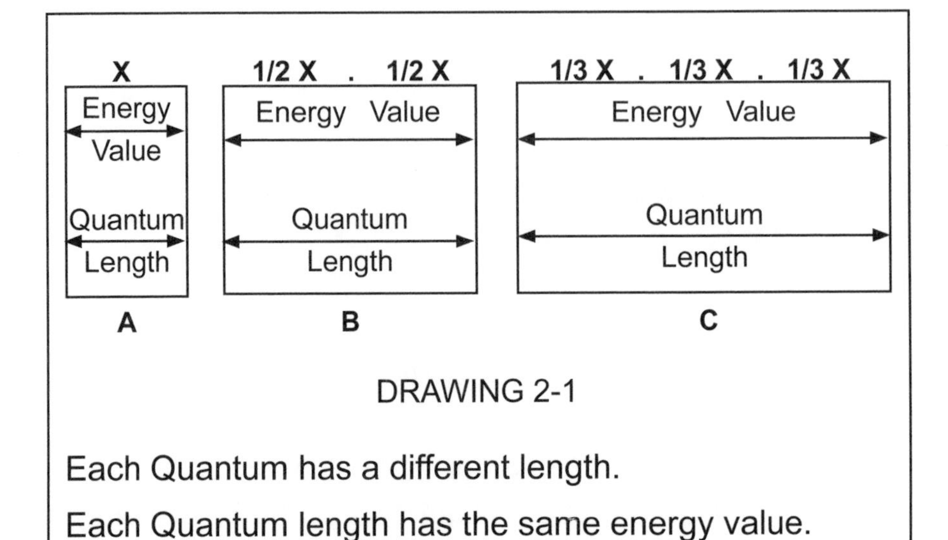

DRAWING 2-1

Each Quantum has a different length.

Each Quantum length has the same energy value.

From the above we can say – a short Quantum as "A" above has all its energy in a short length. A long Quantum as "C" above has all its energy in a longer length. All three lengths "A", "B", "C", each have the same energy value.

There are some other items that need to be considered about quanta formation and its movement. Movement from its place of formation.

At the formation sites - stars, galaxies, our sun, etc. There can be a large number of quanta formed, but just because a larger number are formed does not mean that they are going to move to a destination. What is it that causes the quanta to move to a destination? There are several factors.

It is the temperature of the atom at the time the quanta are formed. The higher the temperature the more electromagnetic momentum the quanta will have. This means that these quanta will travel farther. The inter-atomic forming time of the quantum and the quantum length stay the same. The quanta from stars, galaxies and other electromagnetic sources may or may not reach us here on Earth. It depends on the quanta electromagnetic momentum and the resonance acceptance of Earth. (SEE CHAPTER 13).

ELECTROMAGNETIC MOMENTUM

Explained with today's science terminology. Example: A radio transmitter, transmitting on a certain frequency with an output power of 10 watts – now, increase the output power to 100 watts – the radio signal will travel a further distance. The frequency is still the same. The internal power of the transmitter increased.

CLASSICAL PHYSICS AND ELECTROMAGNETIC PHYSICS

In Classical Physics the speed of an object determines how soon an object gets to its destination. The length of the object has nothing to do with the objects speed. For Example: Assume we have an object 10 meters long moving horizontal at 48 km/hr and another object 5 meters long moving horizontal at 96 km/hr. The destination for each is 300 km distance. The object with the greater speed will get to the destination first. Length of object has nothing to do with speed.

In classical physics, quanta are called wavelengths. The visual wavelengths are usually give in nanometers - nm- which is 10^{-9} meters.

With The New Quantum Theory there are no waves. Now there are packets of energy.

As I mention earlier I use the word quantum instead of wave because what we really have by Quantum Theory are packets of energy; not waves.

In drawing 2-2 we see that the red quantum is longer in length than the violet quantum length. The inter-atomic forming speed of the red quantum is slower than that of the violet quantum which has a faster inter-atomic forming speed.

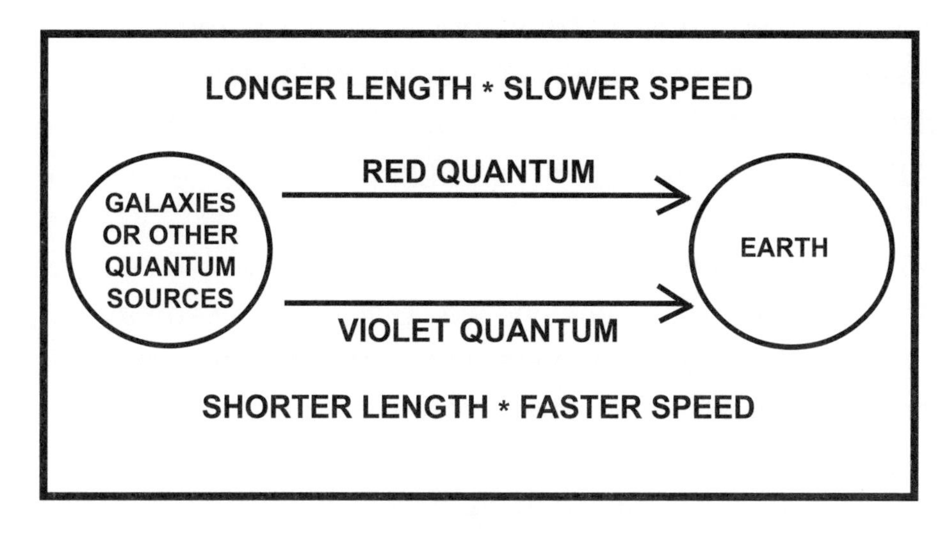

DRAWING 2-2

Will the red quantum get to Earth first or will the violet quantum get to Earth first? What about speed of the quantum? What about length of the quantum?

In Chapter 4 we will find the answers in SPECTROSCOPY and RED SHIFT.

Inter-Atomic Quantum forming time is the electromagnetic time required to produce One Quantum of Electromagnetic Spectrum Energy. This time is different for each quantum. This Electromagnetic time has Electromagnetic distance - not mass distance, as there is no mass.

Atomic time based on Caesium 133 atom and Inter-Atomic Time that I write about in this book are alike in some ways.

Both are Inter-Atomic actions. Atomic Time is a frequency based measurement. My Inter-Atomic forming time is the electromagnetic time required in atomic seconds to form One Quantum of Electromagnetic Energy.

Mass distance is what we have with Ephemers Time. (See Chapter 5) At the present time we do not have a separate Electromagnetic measurement scale with respect to quanta formation time, length, speed, etc.

Scientist realized this and were looking for some form of time that was stable and with more precision. Atomic time came to the rescue. See Chapter 5.

Something I wish to bring out in this book other than the title subject "ELECTROMAGNETIC SPECTRUM -NEW - MANY SPEEDS THEORY" is that there are other avenues of scientific thoughts that are beginning to emerge from the minds of scientists.

The Electromagnetic Spectrum is an entirely separate system from the classical system of physics and the planetary sciences. The Electromagnetic system is a result of action within the atom. The terms I use in this book, Inter-Atomic Quantum Forming Time, Electromagnetic Length, Electromagnetic Speed, Electromagnetic Distance and Electromagnetic Momentum are a result of actions within the atom.

The Inter-Atomic Quantum Forming Time will determine the Electromagnetic Length of the Quantum and the Electromagnetic Speed of the same quantum.

The Electromagnetic Length of the quantum plus the Electromagnetic Speed of the same quantum determines its Electromagnetic Travel Distance.

In Electromagnetic physics the Electromagnetic Length of a quantum is part of the Electromagnetic Speed of a quantum in determining its Electromagnetic Distance movement.

NOTES

CHAPTER 3

ELECTROMAGNETIC QUANTA INFORMATION CHART

After you have studied the Drawing 3-1 to understand it, you will see how quantums of different quantum lengths, all starting at the same zero-point, end. After the first quantum of R, which is R 10, which arrives at the recording instrument before the first quantums of O, Y, G, B, I and V arrive to the recording instrument.

In fact, none of the shorter length quantums will ever catch up with the longer length quantums that they started with.

Also, notice that the energy value of each electromagnetic spectrum (EMS) energy quantum is always the same. Short quantum length or long quantum length.

For example: R to R1, R1 to R2, R2 to R3, etc. are all the same energy value and the energy value R to R1 and the energy value V to V1 are the same. It is the length that changes, not the energy value. As stated earlier, each electromagnetic spectrum (EMS)quantum is a quantum of a certain length and of equal energy for its entire length.

The electromagnetic spectrum (EMS) quanta do not have a self-producing back-flux resistance by their movement. A Mass has a back-flux resistance by its movement. Volkmann Effect. (Chapter 9).

Because the electromagnetic spectrum quanta do not produce a back-flux resistance is the reason electromagnetic spectrum (EMS) quanta can move vast distances. The electromagnetic spectrum (EMS) quanta can move on and on and on.

NOTES

For Drawing 3-1 please see following fold-out page.

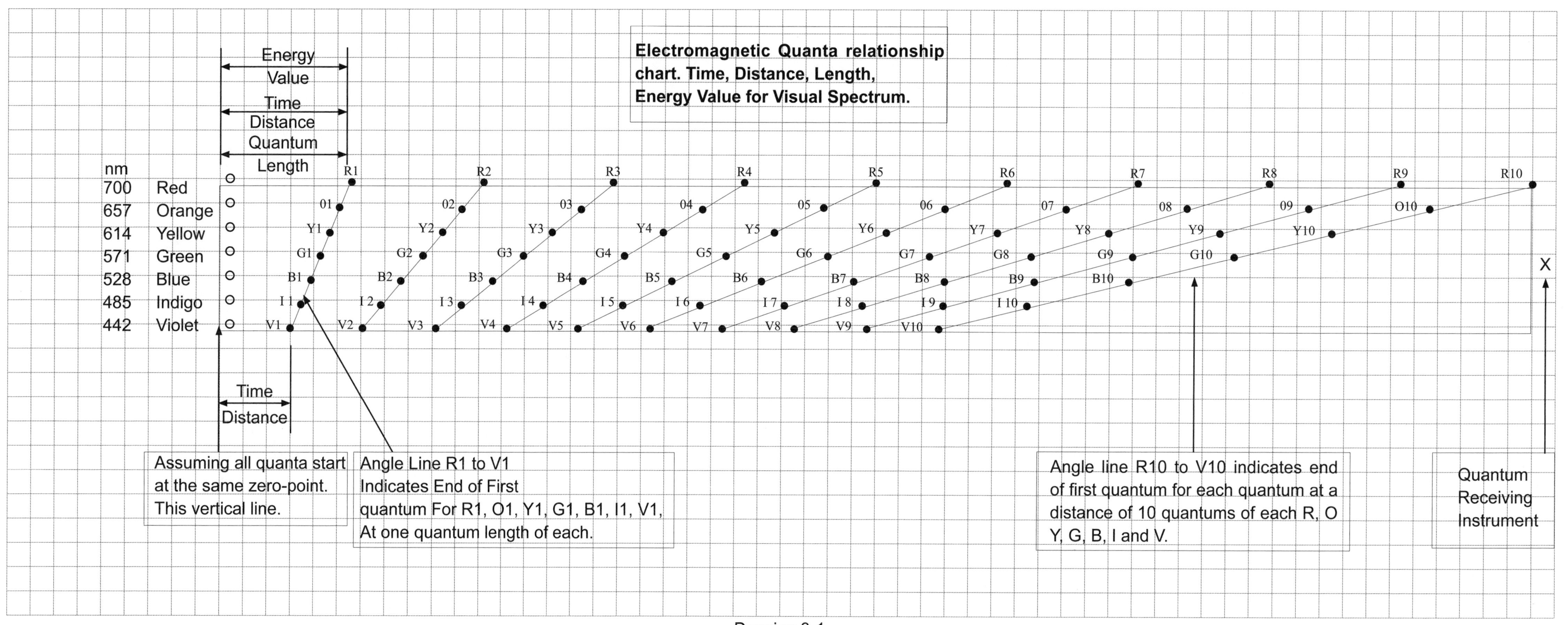

Electromagnetic Quanta relationship chart. Time, Distance, Length, Energy Value for Visual Spectrum.
Energy Value
Time Distance
Quantum Length
nm
700 Red
657 Orange
614 Yellow
571 Green
528 Blue
485 Indigo
442 Violet
R1
R2
R3
R4
R5
R6
R7
R8
R9
R10
X
Time Distance
Assuming all quanta start at the same zero-point. This vertical line.
Angle Line R1 to V1 Indicates End of First quantum For R1, O1, Y1, G1, B1, I1, V1, At one quantum length of each.
Angle line R10 to V10 indicates end of first quantum for each quantum at a distance of 10 quantums of each R, O Y, G, B, I and V.
Quantum Receiving Instrument

Drawing 3-1

CHAPTER 4

SPECTROSCOPY AND RED SHIFT

Now we will look to the science of SPECTROSCOPY and the phenomena of RED SHIFT for more proof of my theory of many speeds of the quanta of the electromagnetic spectrum.

The science called spectroscopy deals with analysis of the electromagnetic spectrum.

Red Shift: Light or other electromagnetic spectrum radiation movement toward the longer quantum length of the electromagnetic spectrum. No change in quantum length or quantum energy value.

Atomic spectra (quanta) are believed to be formed by the movement of electrons in the atom from one energy level to another energy level. The quanta formed by the above action are analyzed by spectroscopic instruments.

Astronomy uses this science in the study of visual light from the stars, galaxies and other sources of electromagnetic radiation. The light spectra are examined to determine what elements make-up the light source, red shift and other information. The spectra of elements here on earth are use as a reference.

In the study of light spectra another feature was noticed on some of the light spectra from distance galaxies. The spectra of these galaxies was shifted toward the red end of the light spectrum. This is called RED SHIFT. Some astronomers believe that this indicates that these galaxies are moving away from earth at a high speed and from this they believe that the universe is expanding very rapidly. This is based on the theory that all spectra of the electromagnetic spectrum have the same speed. My theory does not support this.

There is very little red shift for close distances, but the red shift will start to showup as the distances increase. The more distance, the more red shift. The red shift shows us that the length of the quantum and its speed determine its arrival at the receiver (Earth).

EXAMPLE: SEE DRAWING 4-1.

"Comparison of Red Quantum and Violet Quantum"
The red quantum is longer in length (than the violet quantum) so; it moves a greater distance with each move it makes, but it moves at a slower speed. The violet quantum is shorter in length (than the red quantum) so; it moves a shorter distance

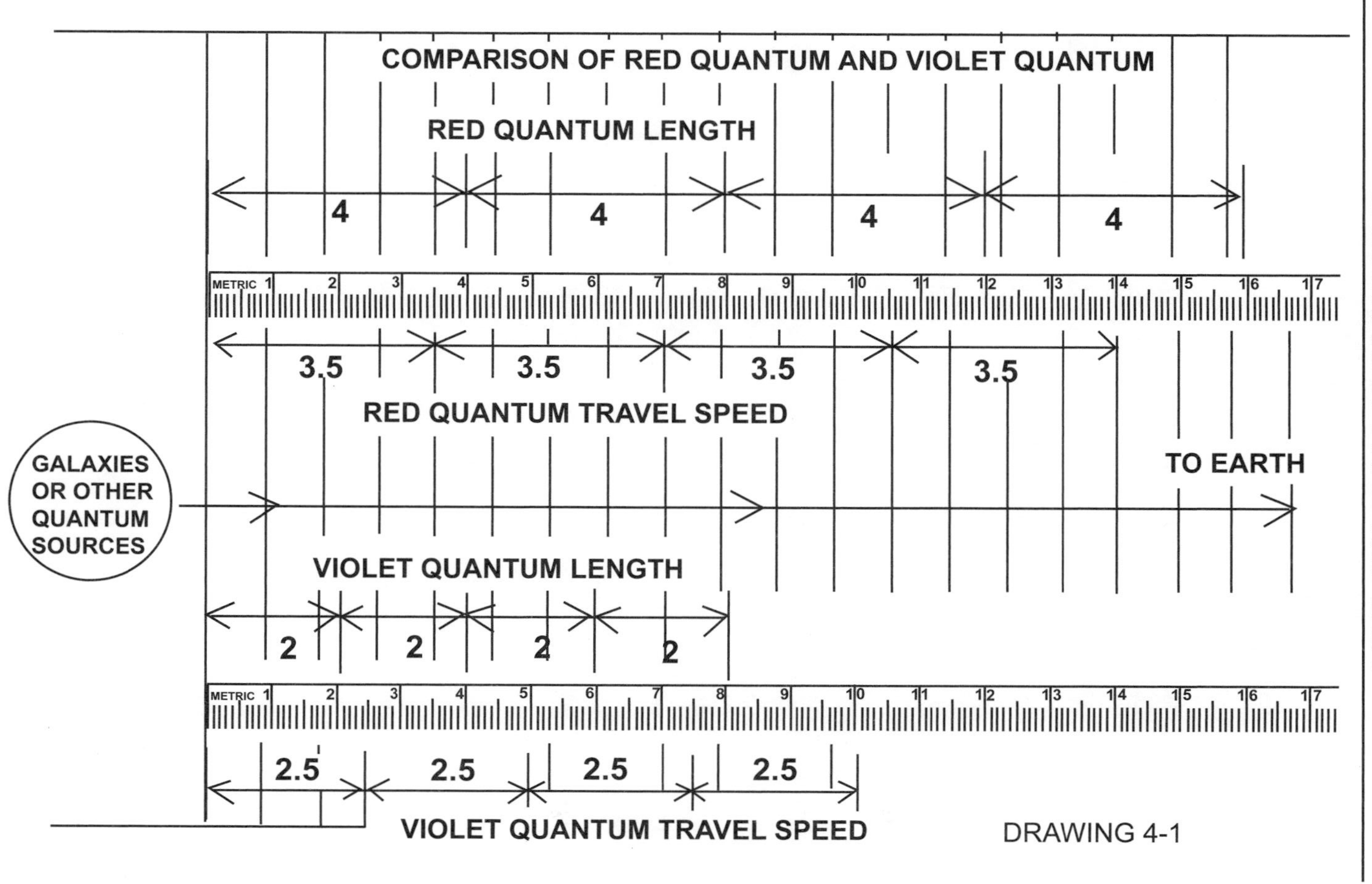
COMPARISON OF RED QUANTUM AND VIOLET QUANTUM
RED QUANTUM LENGTH
4
4
4
4
METRIC
3.5
3.5
3.5
3.5
RED QUANTUM TRAVEL SPEED
GALAXIES OR OTHER QUANTUM SOURCES
TO EARTH
VIOLET QUANTUM LENGTH
2
2
2
2
METRIC
2.5
2.5
2.5
2.5
VIOLET QUANTUM TRAVEL SPEED

DRAWING 4-1

with each move it makes, but it moves at a faster speed.

In DRAWING 4-1 "Comparison of Red Quantum and Violet Quantum": The centimeter distances, quantum lengths and quantum travel speed are used to illustrate and explain the phenomena of how quanta travels in space or vacuum.

In the top part of the drawing (4-1), I show, the red quantum as 4 cm long. The red quantum travel speed is 3.5 cm long (red quantum longer length, slower speed) The longer quantum 4 cm moves in space or vacuum, at a slower speed of 3.5 cm, not at 4 cm.

In the lower part of the drawing (4-1) I show the violet quantum at 2 cm long. The violet quantum travel speed is 2.5 cm long. (Violet quantum shorter length, faster speed). The shorter violet quantum 2 cm moves in space or vacuum at a faster speed of 2.5 cm, not at 2 cm.

Red Quantum Length is determined by its inter-atomic quantum forming time. What is this time? We as scientist do not know yet.

Violet Quantum Length is determined by its inter-atomic quantum forming time. What is this time? We as scientist do not know yet.

Red quantum and violet quantum have different

electromagnetic lengths and different inter-atomic forming times.

All quantums have different inter-atomic formng times and different electromagnetic lengths. Please refer to drawing 4-1. At the top there are 4 red quantum lengths, each quantum has a length of 4 units of length, electromagnetic units of length. This makes a total of 16 electromagnetic units of distance. Each red quantum of 4 electromagnetic units of distance, has a speed rate of 3.5 electromagnetic speed units. (Longer quantum - slower speed).

Each red quantum of 4 units of electromagnetic distance plus 3.5 units of electromagnetic speed units equal 7.5 red quantum units or electromagnetic distance for each red quantum move.

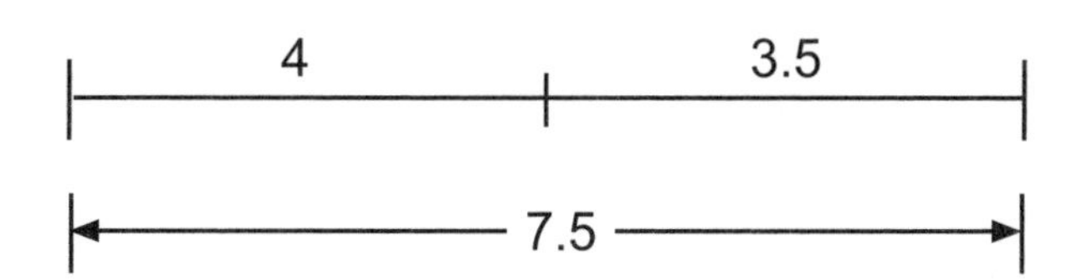

Electromagnetic Distance

Violet Quantum - at the lower part of Drawing 4-1 there are 4 violet quantum lengths, each quantum has a length of 2 units of length, electromagnetic units of length. This makes a total of 8 units of electromagnetic units of distance. Each violet quantum of 2 electromatnetic distance units has a speed rate of 2.5 electromagnetic speed units. (Shorter quantum - faster speed). Each violet quantum of 2 units of electromagnetic distance plus 2.5 units of electromagnetic speed units equal 4.5 violet quantum units of electromagnetic distance for each violet quantum move.

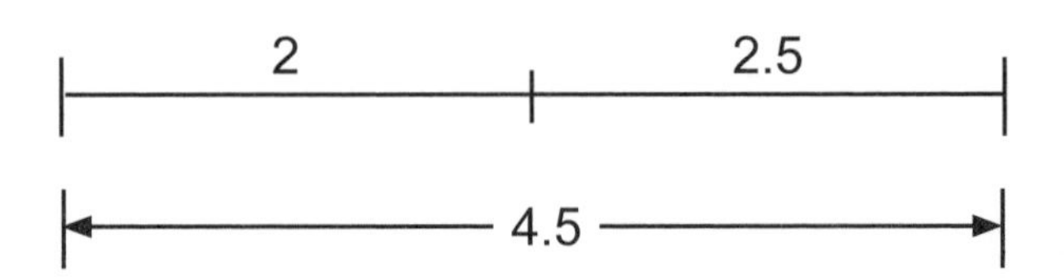

Electromagnetic Distance

From the information above we can see that a longer quantum [red] will get to its destination before a shorter quantum (violet). See drawing 4-1 both leaving the sender at the same time.

On Drawing 4-1 and see above. The red quantum travels 7.5 units of electromagnetic length with each move.
On Drawing 4-1 and see above. The violet quantum travels 4.5 units of electromagnetic length with each move.

In Drawing 4-2 we can see what is called red shift. The inter-atomic quantum forming time and the inter-atomic forming speed is different for each quantum. This is the cause of the red shift effect.

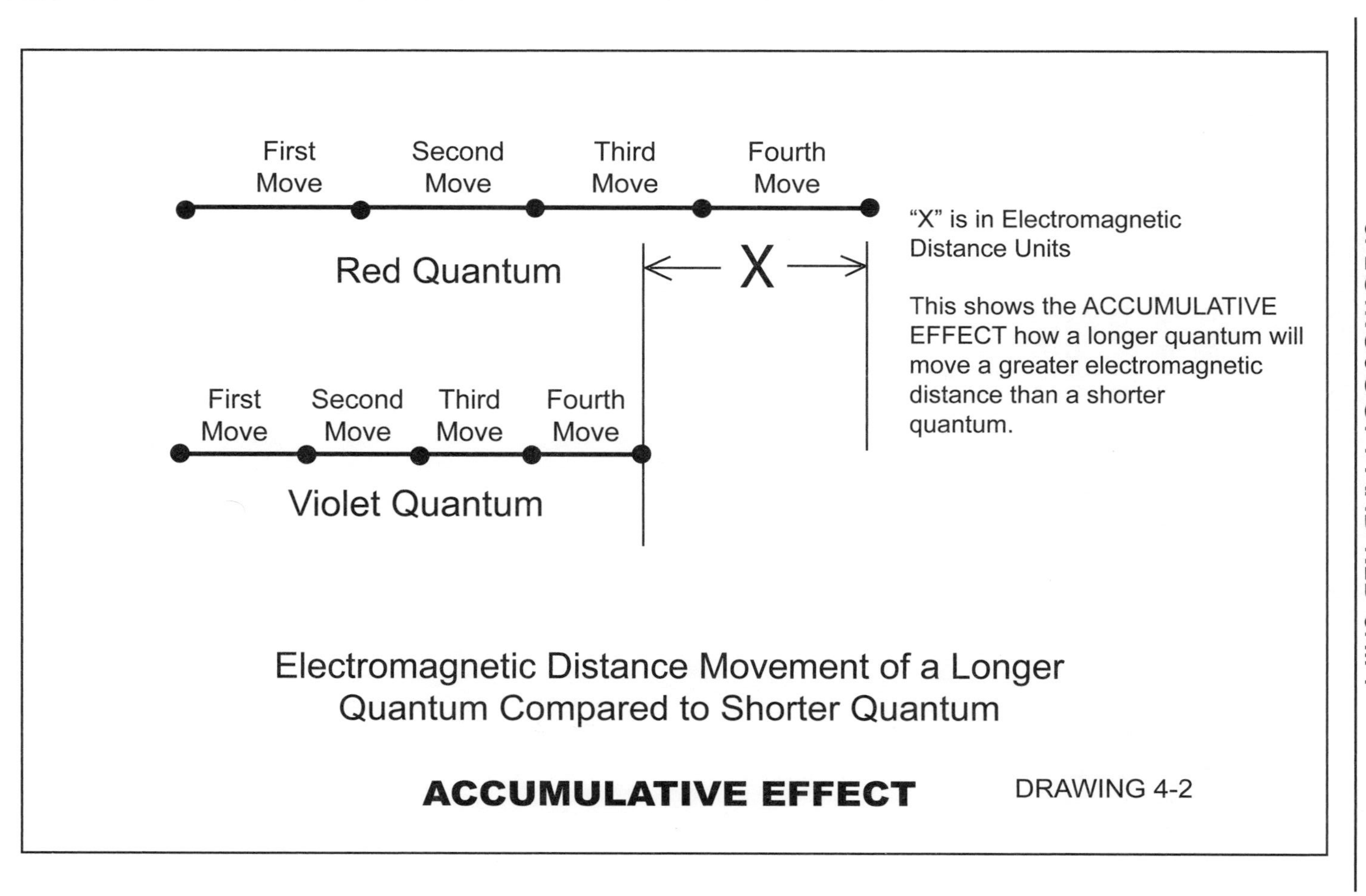
First Move
Second Move
Third Move
Fourth Move
Red Quantum
X
"X" is in Electromagnetic Distance Units
This shows the ACCUMULATIVE EFFECT how a longer quantum will move a greater electromagnetic distance than a shorter quantum.
First Move
Second Move
Third Move
Fourth Move
Violet Quantum
Electromagnetic Distance Movement of a Longer Quantum Compared to Shorter Quantum
ACCUMULATIVE EFFECT
DRAWING 4-2

Now assume that a red quantum and a violet quantum leave the sender at the same time. The electromagnetic distance of the red quantum and the violet quantum is so minute that the ACCUMULATIVE EFFECT (drawing 4-2) will be very small at short distances, but will become noticeable as the distance from sender to receiver increases. This causes the red shift to increase with distance.

SO; THERE YOU HAVE IT. THIS IS MY THEORY FOR A MANY SPEEDS ELECTROMAGNETIC SPECTRUM.

THE RED SHIFT INFORMATION GIVE ABOVE PROVES THAT THE ELECTROMAGNETIC SPECTRUM IS MADE-UP OF MANY SPEEDS NOT JUST ONE.

NOTES

CHAPTER 5

DISTANCE AND TIME, ARE THEY THE SAME?

Are distance and time the same or are they different? Now, let me explain my views on distance and time.

On distance, we have fixed distance and moving distance. Fixed distance is measured distance here on earth. That is, a person measures distance on earth with his measuring tools.

If the distance on earth between x and y is 160 meters, the distance is a fixed distance. A person can walk or run the distance and it is still 160 meters. The earth rotates on its axis and in its orbit about the sun, the distance between x and y is still the same 160 meters, this is a fixed distance.

Moving distances are the moving earth, planets, sun, stars, galaxies, etc. These objects are constantly changing distances among themselves.

Most actions, partcle movement, chemical reactions, clock time, life, etc., are usually distance-time movement. That is,

they are measured to the earth's rotation on its axis or earth's orbit around the sun. This is called ephemers time.
Example: If a person lives 72 years – time – what we mean is he has lived a distance of 72 orbits – distance – of the earth around the sun.

Are time and distance the same, or just different names for the same phenomena?

On fixed time, the meter length, here on earth, is still the same distance after 72 orbits around the sun. The time or distance involved did not change the meter length on earth.

For Mass movement measurements we use fixed earth distances.

There is also electromagnetic spectrum movement distances. Here we use a fixed earth distance for measurement of an electromagnetic spectrum quantum distance. This is not the best way, but at the present time we have no other way.

Mass With Energy and the electromagnetic spectrum are two entirely different systems.

A fixed distance here on earth can be measured with our measuring devices, fixed distances involve Mass. When a fixed distance involves movement we usually use earth's rotation on its axis and/or rotation around the sun as a reference. This is ephemers time.

However, when there is movement of electromagnetic spectrum energy, we come upon new problems. Are we going to use fixed distance measuring methods to measure fast moving electromagnetic spectrum energy? That is what we do now.

How does one go about measuring electromagnetic energy distance movement without using fixed distance methods?

The problem is that we have moving electromagnetic spectrum energy distances and with this we have to deal with an uncertainty. That is everything is moving at different speeds, at different distances, and at different distance-times.

The German astronomer and mathematician Johannes Kepler (1571-1636) discovered that the planets rotate around our sun in elliptical orbits and they sweep out equal areas in equal times. This was a very important discovery.

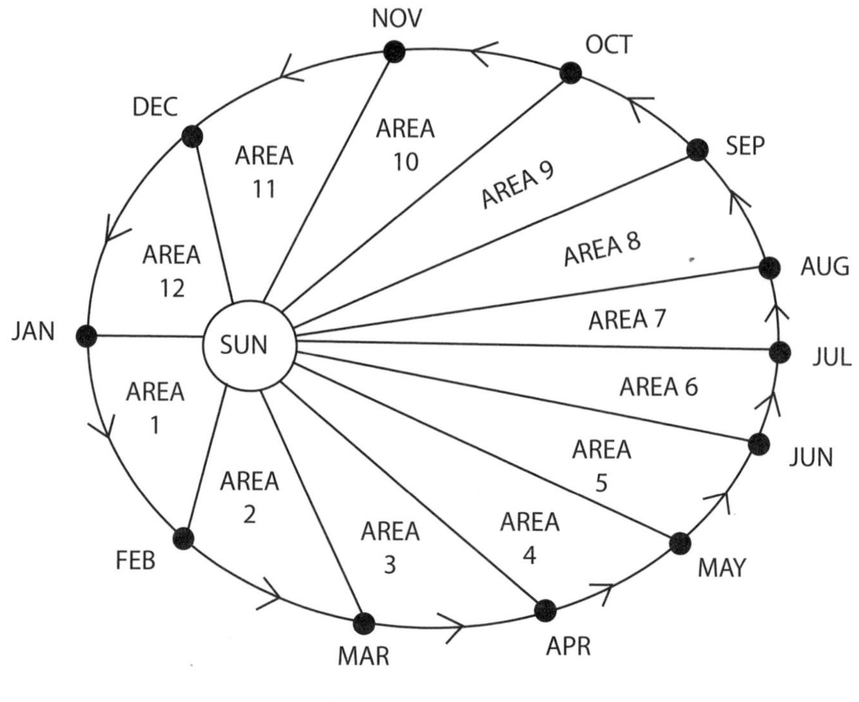

DRAWING 5-1

Here we see the earth in its elliptical orbit around the sun. Kepler's law states that the earth (and other planets) sweep out equal areas in equal time intervals.

Kepler's theory is a very interesting, "equal areas in equal times" Drawing 5-1.

When the earth is closest to the sun as it is in December, January and February it moves faster in its orbit, moving more distance for each interval. During June, July and August when the earth is farthest from the sun it moves slower in its orbit, moving less distance during each interval. As the earth travels in its ecliptic orbit about the sun we can see that its speed is different for each "time" second or orbit travel. So; from all this we can say for earth.

1. Less distance to the sun the faster its orbit speed.
2. More distance to the sun the slower its orbit speed.

And: equal areas in equal times; also; to me this means that the equal areas are equivalent to equal energy received by earth from the sun in equal orbital areas.

There are many categories of time.

1. Ephemers Time: Time measured by the orbital movements of the earth, the moon and the planets. This ephemers time has mass distance since it involves movement of mass, for everyday time measurements, axial rotation of earth is used.

Ephemers Time is not an accurate time for science. This is due to many factors. Axial rotation of the earth, inclination of the ecliptic, eliptical path of the earth around the sun, irregular rotation variations of earth around our sun, gradual slowing of earth's rotational speed, etc. all this made for a varying second "time".

Scientists realized this and were looking for some form of time that was stable and with more precision this is where atomic time came to the rescue.

ATOMIC TIME

In 1967 the general conference of weights and measures

adopted a new definition of the SI Second based on certain structure transitions in the caesium 133 atom.

SI = International System of Units

Inter-atomic quantum forming time is the time required to produce one quantum of electromagnetic spectrum energy. This time is different for each quantum. This time has electromagnetic distance – not mass distance.

Atomic time based on caesium 133 atom and inter-atomic time that I write about in this book are alike in some ways. Both are inter-atomic actions. Atomic time is a frequency based measurement. My inter-atomic forming time is the electromagnetic time required in atomic seconds to form one quantom of electromagnetic energy.

Mass distance is what we have with ephemers time.

At the present time we do not have a separate electromagnetic measurement scale. Why all this about time? In this book ELECTROMAGNETIC SPECTRUM – NEW – MANY SPEEDS THEORY I use time for the terms inter-atomic forming time, quantum time and electromagnetic time. These times are a form of atomic time, not ephemers time.

For more on this subject see Electromagnetic Physics Chapter 2.

CHAPTER 6

ELECTROMAGNETIC SPECTRUM AS IT IS NOW

Present information on the electromagnetic spectrum shows that the electromagnetic spectrum is composed of many frequencies that are sinusoidal wave form, that is the wave starts at zero, goes to peak, back to zero, back to peak, back to zero. This looks like a type of electric sine wave produced by a coil of wire rotating in a magnetic field.

In 1775 the Danish Astronomer, Roemer, determined the speed of light. His method determined how long it took light to travel the major axis of the earth's orbit.

The sun's spectrum as seen on earth peaks around 500nm. The human eye for scotopic vision (low light and night vision) also peaks around 500nm.

The wavelength of the light – by classical physics – seen by Roemer must have been mostly around 500nm, so it would seem that the light he saw was in this 500 nm region. This being based on the speed of light being around 3.0×10^8

meters/second.

In the following two Drawings 6-1 and 6-2, one can see the way the electromagnetic system is now. Any number of cycles per second. 1000 Hz/sec., 10,000 Hz/sec., 100,000 Hz/sec., etc. The second is the time base.

The system is based on the time unit of one second. The number of wavelengths per second.

THE WAY IT IS NOW. ANY NUMBER OF Hz / Sec.

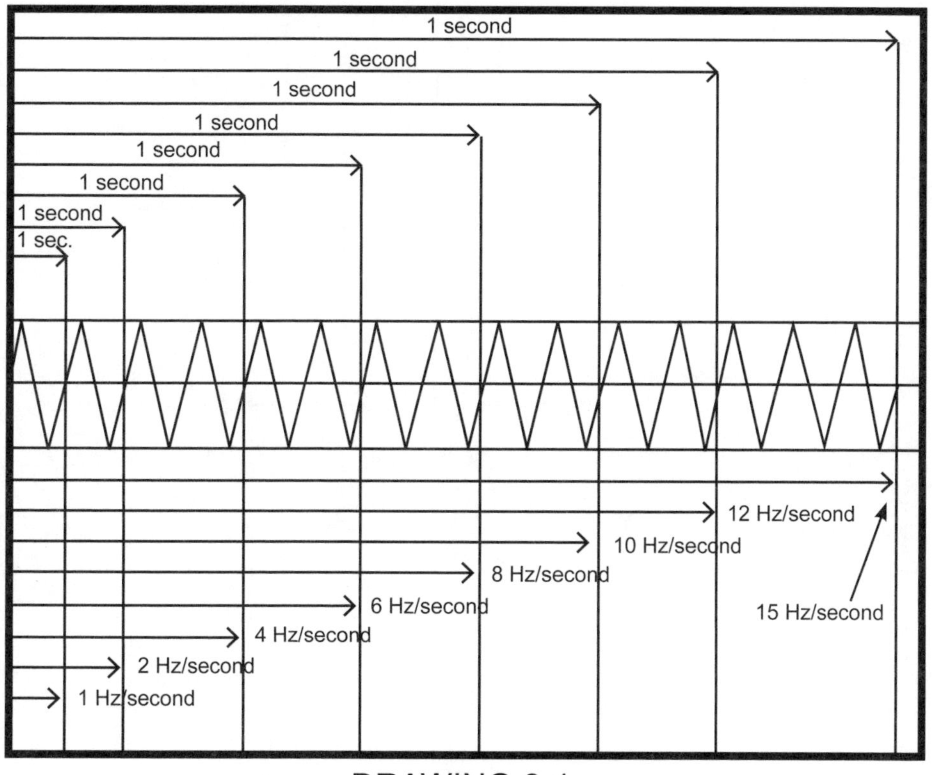

DRAWING 6-1

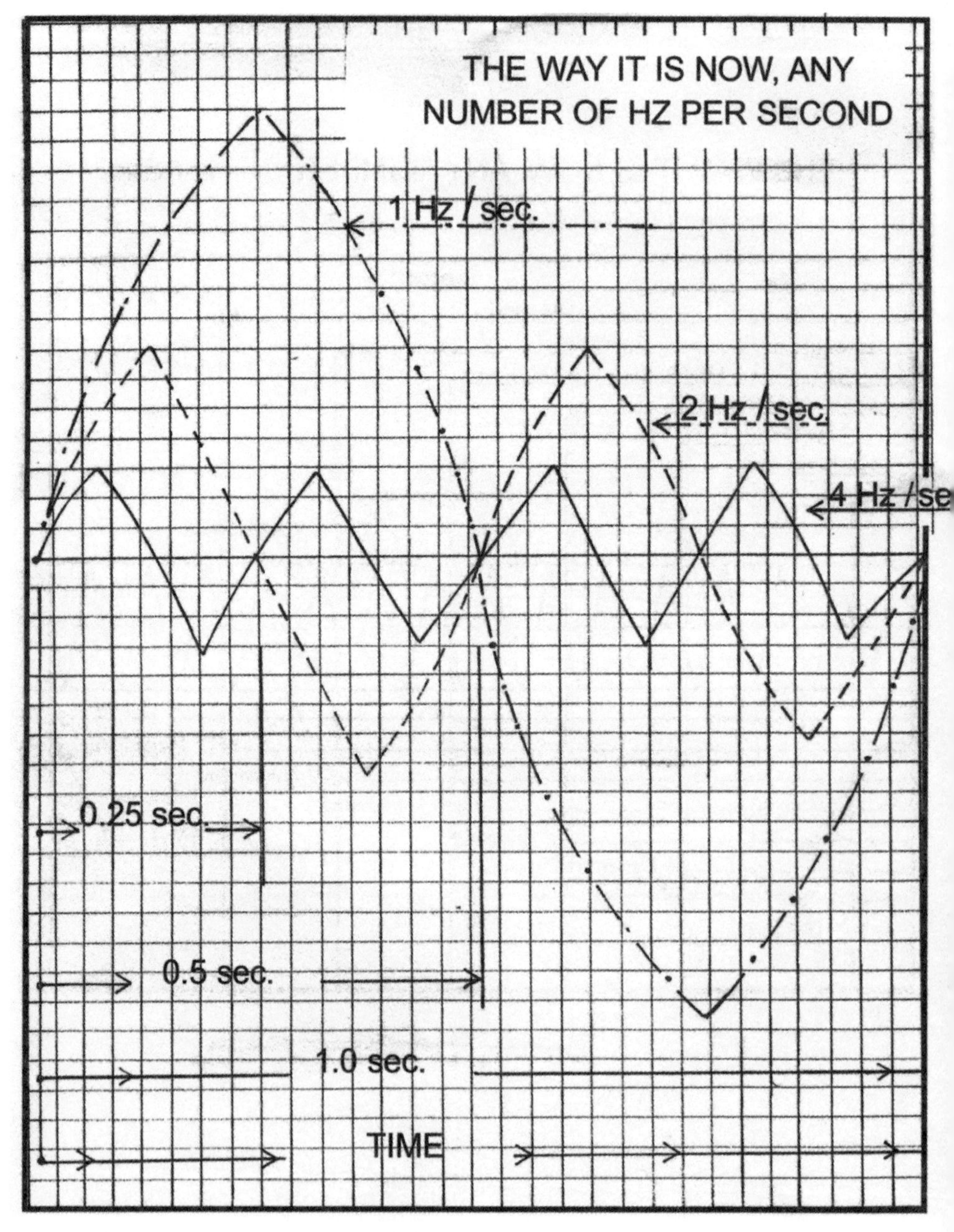
THE WAY IT IS NOW, ANY
NUMBER OF HZ PER SECOND
1 Hz / sec.
2 Hz / sec.
4 Hz / se
0.25 sec.
0.5 sec.
1.0 sec.
TIME

DRAWING 6-2

This method of using the SI second as a time base with frequency is fine for many uses.

However, with the advancing atomic age where the SI second is a very long time compared to inter-atomic actions.

We need to measure atomic actions (quantum forming time and other inter-atomic actions) as parts of the SI second, not how many atomic actions per second.

NOTES

CHAPTER 7

ELECTROMAGNETIC SPECTRUM ENERGY FROM VAST DISTANCES

The movement of an electromagnetic spectrum quantum does not produce a back-flux resistance or have contraction of length, because it has no mass, just energy.
(SEE CHAPTERS 9 AND 10.)

The electromagnetic quanta can move on and on and on. Since electromagnetic energy needds no carrier and produces no back-flux resistance -Volkmann Effect. (Chapter 9).

This is the reason electromagnetic quanta can move vast distances. Examples: light energy, x-ray energy, radio energy, etc. All from vast distances.

NOTES

CHAPTER 8

ELECTROMAGNETIC SPECTRUM QUANTUM SPEED IN A TRANSPARENT MEDIUM

Speed of an electromagnetic spectrum quantum as it goes through a piece of clear glass. (Medium)

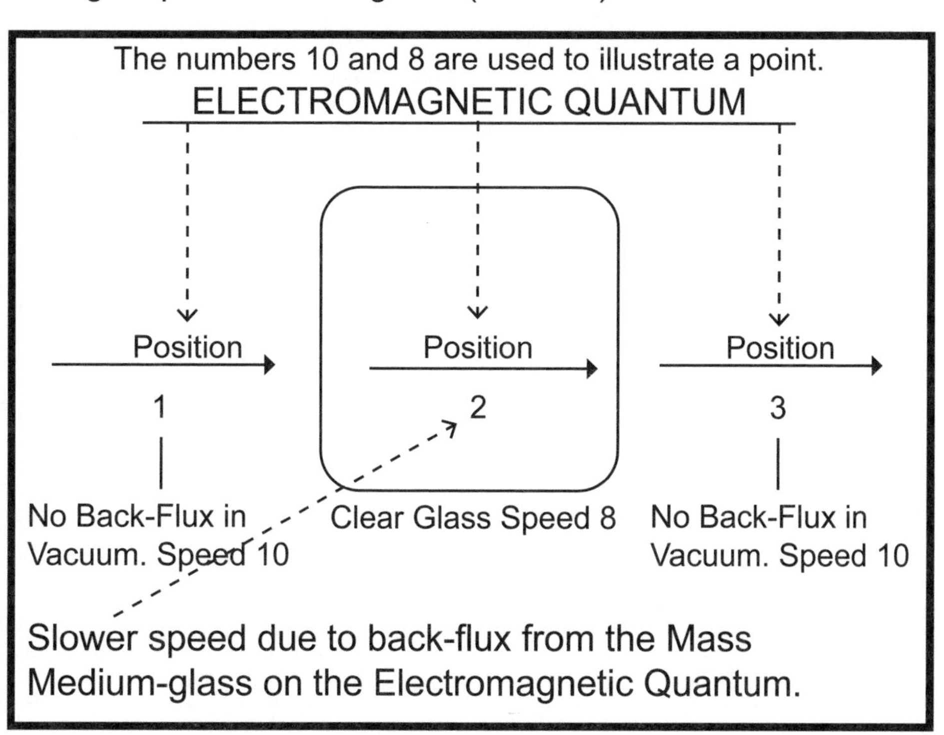

DRAWING 8-1

NOTES

CHAPTER 9

VOLKMANN EFFECT
MOVEMENT OF "MASS WITH ENERGY" AND ITS SELF-INDUCED BACK-FLUX RESISTANCE
– AND –
WILL AN OBJECT – A MASS WITH ENERGY – LAUNCHED INTO SPACE EVER STOP?

A MASS launched in the direction of earth's rotation will have the earth's west to east rotational energy plus the launch energy.

Since the space medium is not very conductive to receiving the two energies, mentioned above, the moving mass could go in space forever.

However, the moving MASS WILL NOT go traveling forever. There is another factor to consider. That is the VOLKMANN EFFECT. It is known as Back-Flux Resistance of a moving MASS in Galactic Space Energies. This will finally bring the moving MASS to a stop.

All of the launched energy and energy from earth's rotation, west to east will have been canceled by the self – produced Back-Flux Resistance, little by little as it moves through space, to the point where there is no motion energy left.

A MASS moving in Galactic Space Energies will have resistance to moving. In this case, the resistance to moving is caused by a self-induced back-flux of energy produced by the moving MASS in the Galactic Space Energies. "Volkmann Effect".

In outer space, that is our galactic space and extra galactic space, it has been shown that there are many forms of Electromagnetic Spectrum Energies. For example: cosmic, x-ray, ultra-violet, visual, radio and others.

In this case the Back-Flux Energy causes the Mass movement to slow. The Back-Flux Energy slowly neutralizes the forward Mass Movement Energy.

Now what happens to the Mass when its movement energy becomes zero?

What I think will happen is the "dead" mass will become the captive of the nearest space energy flux.

AS A RESULT OF A MASS MOVEMENT IN THE GALACTIC SPACE ENERGIES, A BACK-FLUX ENERGY IS DEVELOPED THAT OPPOSES MOVEMENT OF THE MASS THAT CAUSED THE BACK-FLUX ENERGY TO BE PRODUCED. THIS EFFECT IS CALLED THE "VOLKMANN EFFECT."

This is similar to electrical phenomenon known as Lenz's Law. This is where an electric current flow in a wire produces its own back-flux resistance to the electric current flow. This current back-flux resistance is in addition to the wire resistance to current flow.

NOTES

CHAPTER 10

CONTRACTION OF LENGTH AND SHRINKAGE OF A MASS

Note: This chapter can be very confusing, so read it with this in mind and pay close attention to details.

Developed theories on the properties of moving objects, such as the length of an object becomes shorter the faster it moves.

After studying these contraction theories, I did not think that they were correct, so I developed my own theory as to what I think takes place.

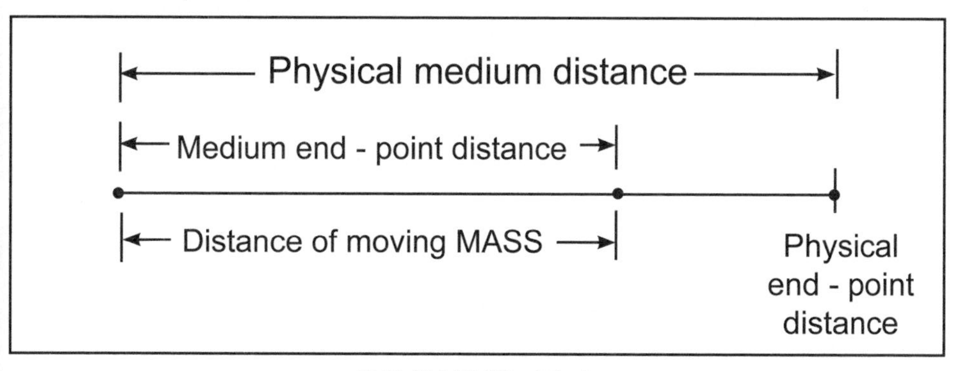

DRAWING 10-1

It is not the moving Mass that shrinks, it is that the medium end-point distance is created, it being shorter with respect to the physical distance of the medium.

As you read on, you will understand this better.

The moving Mass will not go to the physical distance end-point as it does not exist for any moving Mass.
The moving Mass will not reach the physical end-point because the energy level of the moving Mass is before the physcial end-point is reached.

The movin Mass energy level is at the end-point of the shortened physical end-point which is the medium end-point distance.

The shorter medium end-point exists for any moving Mass.

Where is the moving Mass medium end-point?

It is always before the physical medium end-point due to contraction of the medium distance.

There is no contraction of the moving Mass or of the medium Mass. There is only a contraction of distance.

The medium Mass end-point distance is at a lower energy level than the end-point energy level of the physical end-point distance.

Where is the missing energy?

The difference in energy level between the physical end-point distance energy level and the medium end-point distance energy level is the resistance of the movement of the moving Mass in the medium Mass.

Remember, it is the distance that has contracted, not the moving Mass or the medium Mass.

Now that I have explained my theory on contraction of length of distance of a moving Mass in a medium Mass, we need to see the subject in more detail.

The physical distance end-point will now be called the imaginary end-point (the non-moving Mass end-point).

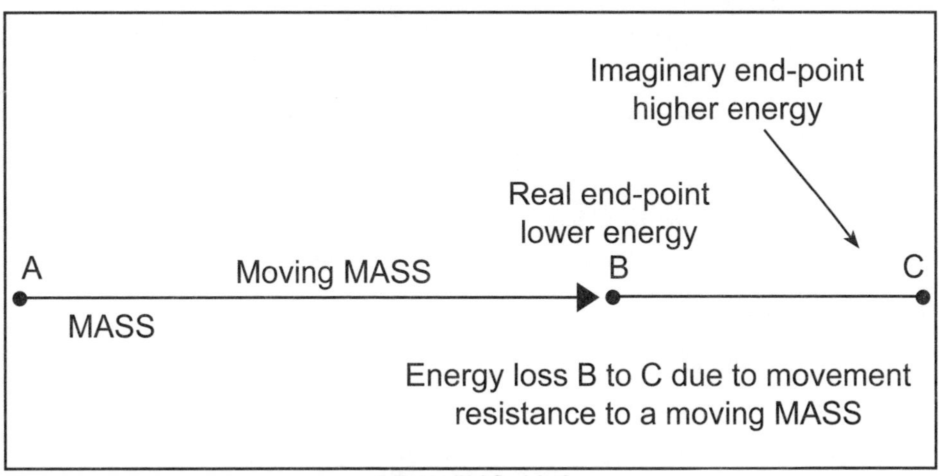

DRAWING 10-2

The medium distance end-point will now be called the real end-point (the moving Mass end-point).

There are two distance end-points. The imaginary distance end-point, C, and the real distance end-point, B. SEE DRAWING 10-2.

The non-moving Mass distance end-point is the imaginary distance end-point, C, in reality it does not exist. It is the distance end-point a Mass would go to if there were no resistance to the movement of the moving Mass.

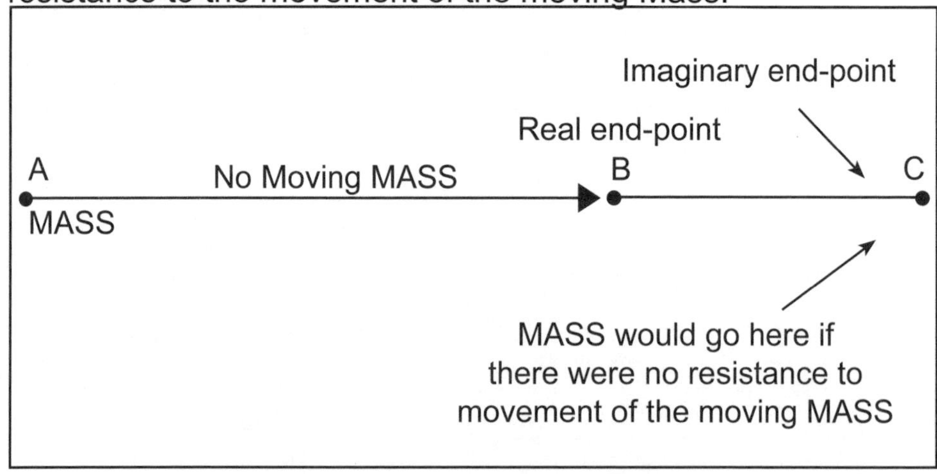

DRAWING 10-3

The moving Mass distance end-point is the real distance end-point, B. In reality it does exist. It is the distance end-point a Mass does go to, because there is movement resistance to the moving Mass.

The moving Mass will go to the real end-point, because the

real end-point is its energy level end-point.

IF THERE IS NO MOVEMENT OF A MASS, THEN THE IMAGINARY DISTANCE END-POINT DOES NOT EXIST, IT IS ONLY IMAGINARY.

HOWEVER, IF THERE IS MOVEMENT OF A MASS THEN THE IMAGINARY DISTANCE END-POINT DOES EXIST.

HOWEVER, FOR THE MOVING MASS THE IMAGINARY DISTANCE END-POINT DOES NOT EXIST.

ANY MOVING MASS CANNOT REACH ITS FULL POTENTIAL DUE TO MOVEMENT RESISTANCE TO A MOVING MASS.

If you understand what is blocked, then you understand what I am writing about.

In this article I have not included “self-induced back-flux resistance of a moving Mass”, but, it too is part of the moving Mass discussed above. Volkmann Effect.

NOTES

CHAPTER 11

GRAVITY

Now to explain gravity.

Gravity is said to be the attraction force between two or more masses. Gravity is also said to be the cause of weight of a mass.

Are these concepts well explained in today's science?

Let us look and analyze to see what gravity really is.

What we call weight of a mass is not caused by what we call gravity.

Weight of a mass is caused by rotation of the earth which produces the centrifugal force. The faster the earth rotates, the less a mass will weigh.

This is due to the fact that there will be more centrifugal force produced from an earth that is rotating faster. This faster

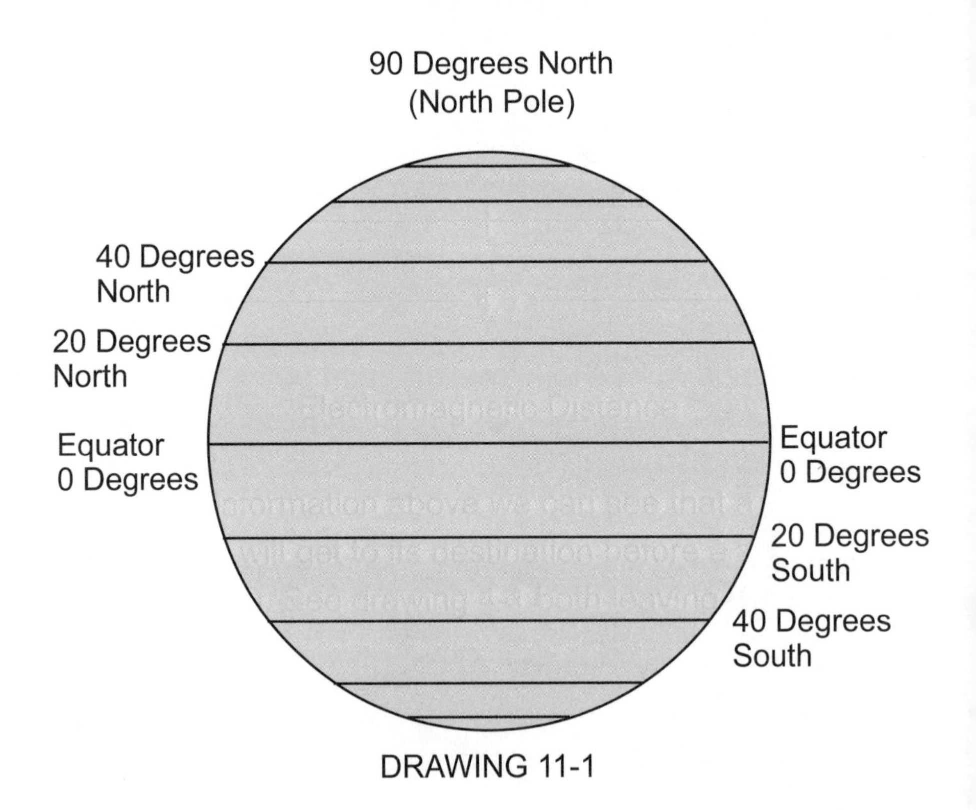
Latitude
90 Degrees North
(North Pole)
40 Degrees
North
20 Degrees
North
Equator
0 Degrees
Equator
0 Degrees
20 Degrees
South
40 Degrees
South

DRAWING 11-1

rotation will cause a mass object to weigh less and escape from earth, if the mass-object is not held to tightly by internal binding forces.

At the equator – zero latitude the earth is larger in circumference than at higher latitudes. So; what this means is that the equator – zero latitude – rotates at a faster speed than higher latitudes.

A 1 kilogram object at a higher latitude will weigh more than 1 kilogram at the equator - 0 latitude. (Same 1 kilogram object from higher latitude used for measurement at equator). This is because at the equator, the earth has a longer circumference (more distance) to rotate in the same time as the higher latitude.

What we call weight is caused by the centrifugal force of the rotating earth. This is not gravity.

Now! Then what is gravity? See page 99.

You remember early, that the definition of gravity is the attracting force between two masses.

If you have not read Chapter 16 then you should stop here and read Chapter 16 so you will understand the information to be presented.

Now! That you have read Chapter 16 we can proceed.

DRAWING 11-2

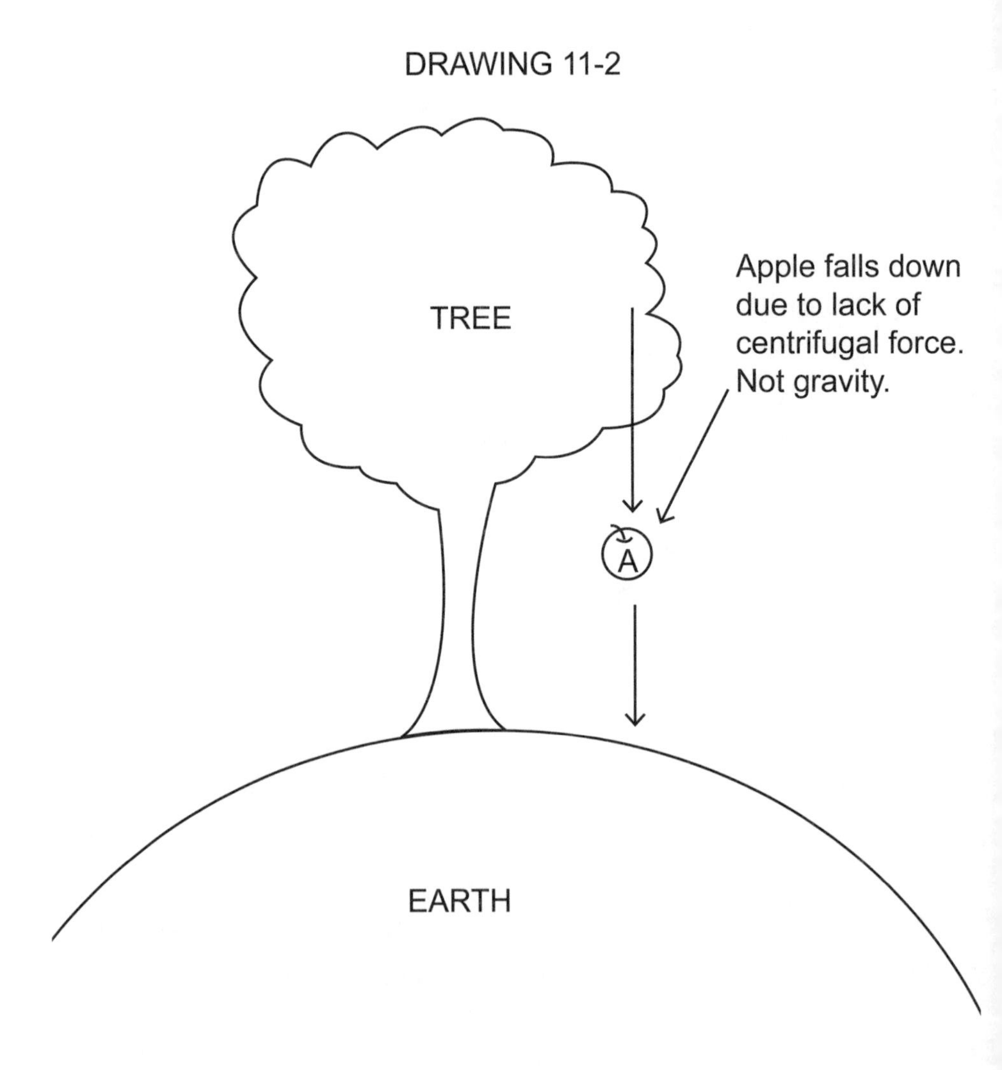
TREE
Apple falls down due to lack of centrifugal force. Not gravity.
A
EARTH

DRAWING 11-3

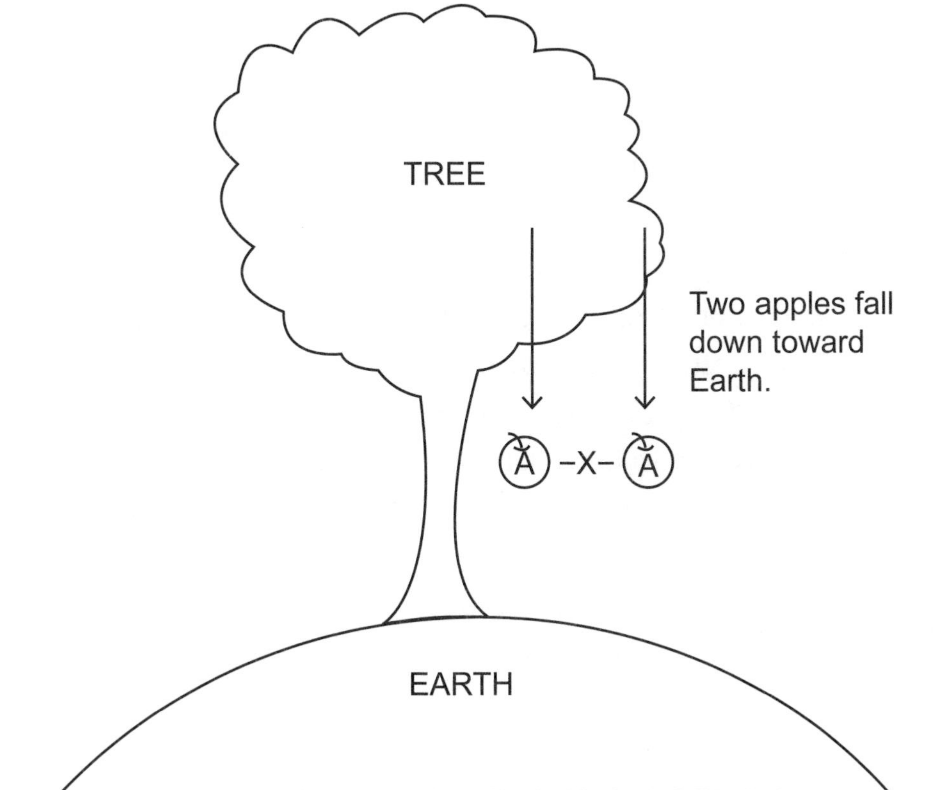

TREE

Two apples fall down toward Earth.

A –X– A

EARTH

THE APPLES FALLING DOWN IS CAUSED BY LACK OF CENTRIFUGAL FORCE.

THE ATTRACTIVE FORCE X BETWEEN THE APPLES IS GRAVITY. BOTH ARE FALLING IN THE SAME DIRECTION.

All mass of the earth is moving in the same direction in its elliptical orbit, so each mass particle has some attraction with every other mass particle.

There is a lot more to this GRAVITY chapter that I have not written.

1. What if there is no movement of any mass – would there still be gravity?

2. Assume we have two spheres, each 1 cm in diameter, of the same mass, (Element) and 1 cm apart, moving in the same direction. The direction East to West, North to South, West to East, South to North; also the centrigugal force and the Coriolis force, will have their effect.

   I do not have the time to make a detail study at the present time.

NOTES

NOTES

CHAPTER 12

SEND – RECEIVE THEORY

Why does light from a distant star come to us on earth?

Why are we able to see distance or receive other electromagnetic spectrum (EMS) energy?

(Please read my theory on ELECTROMAGNETIC SPECTRUM ENERGY FROM VAST DISTANCES), Chapter 7.

Now back to the article. There must be a receiver for this energy before there can be a sender of the energy.

How does the sender know whether there is a receiver or no receiver?

Now let us see how this sending and receiving works.

Assume that there is a star that has a lot of potential energy. The star wishes to get rid of this energy before it explodes or transforms into more Mass. So it sends out weak energy quanta of electromagnetic energy in all directions – seeking receivers to accept this energy. If there are not enough receivers for this energy, the star will start transforming to more Mass. Conversion of energy to Mass.

The weak Electromagnetic Spectrum (EMS) quanta go out seeking receivers for this energy, and they come upon a receiver and the receiver accepts this weak quanta.

Then, this leaves an empty area where the quanta was before. Other Electromagnetic Spectrum (EMS) quanta move into this empty area from the same source. Soon we have a stream of Electromagnetic Spectrum (EMS) quanta from this specific source being accepted by this specific receiver.

A receiver can only receive that amount of energy that it is in "resonance acceptance" with the sender.

The receiver does not send out a signal to the sender, and then the sender sends a signal back to the receiver.

The earth and planets do not have enough Electromagnetic Spectrum (EMS) Energy to send much out to the receivers. The earth and planets are more receivers of EMS Energy than senders of EMS Energy. Unless it happened to be reflective EMS Energy, which is actually EMS Energy from another source.

Another item that I would like to present on the send-receive theory is radiation of energy from the sun.

Does the sun radiate energy equally or almost equally from all of its spherical surface?

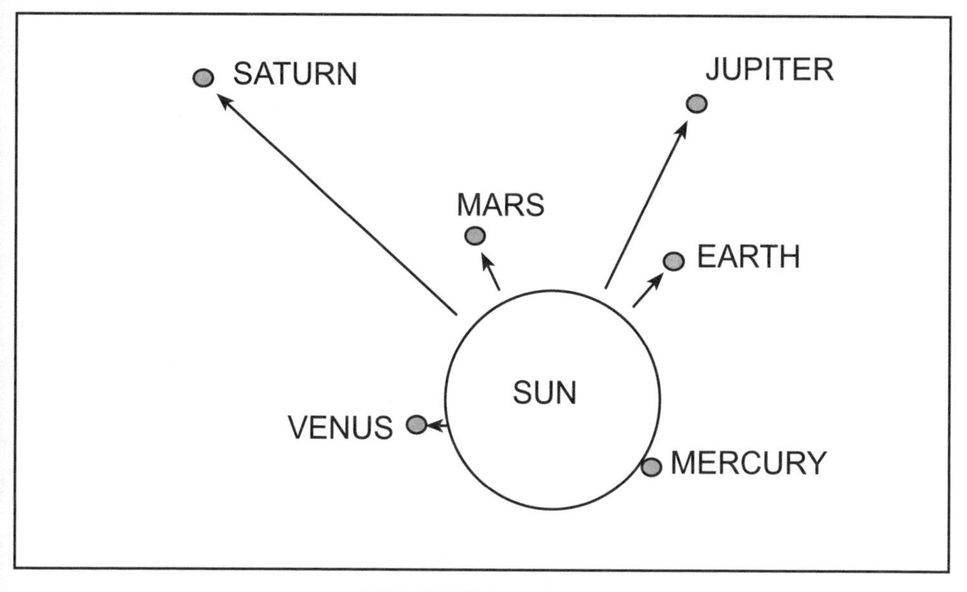

DRAWING 12-1

No it does not radiate energy equally from its spherical surface.

The radiant energy will go to acceptors of the suns radiant energy. The physcial size, distance and degree of resonance acceptance are all factors.

What about the sun's spherical surface areas that do not

have receivers.

No receivers means no sending. There must be a receiver before there can be a sender.

So the area of the sun's spherical surface that do not have receivers for their energy will not be able to send their energy.

In a later paper you will see that the sun has very small amount of attractive energy, but has a lot of repulsion energy.

CHAPTER 13

THEORY OF “RESONANCE ACCEPTANCE”

Everything, that is all Mass is in a state of some degree of “resonance acceptance” with every other Mass.

All Mass that we have here on earth is made of the periodic table of elements.
Some of the man-made elements may or may not exist as part of the larger galactic system.

I am assuming that everything in all the galactic systems “hatched from the same egg”.

This means that everything is related, but this does not mean that everything is the same.

We see and detect only what we are in resonance acceptance with, that is what our 92 elements are in resonance acceptance with.

Earlier I said “everything being hatched from the same

egg". To me this means that everything is related by energy level. Does this mean that everything is at a different energy level, yes. The fact that everything is at different energy levels is the reason that movement from one energy level to another energy level keeps the whole system going.

There is a continual energy exchange going on all the time. For example: Energy from the sun to earth and other planets, light energy from stars, radio energy from seen and unseen places, x-ray energy, cosmic energy, movement of planets.

Why would we on earth receive any of this energy or be part of this great energy exchange?

My answer to this question is "we receive these energies here on earth because of our resonance acceptance" with these received energies".

All the Mass, being all the 92 periodic table elements, are in a varying degree of energy in "resonance acceptance" with every other periodic table element.
No matter where the energy is, it could be on earth or part of a star or galaxy.
Nothing has to be sent to the star, galaxy, sun or supernova, to say, "here we are, send us your energy, light, x-rays, particles, etc." Please read my "Send-Receive Theory". Chapter 12.

No, all the energies, particles, Mass, etc. are all part of the original energy package. Everything has a varying degree of energy.

This energy will transfer only to a "resonance acceptor".

It has nothing to do with the amount of energy that a sender or receiver will accept it has to do with the amount of "resonance acceptance" one has for the other.

Now, I am sure many of you readers would like to have an example of what I am writing about.

For those of you that have some knowledge of electricity, let me explain a phenomenon of energy transfer in radio transmitting.

In a radio transmitter the final circuit before the antenna there is a coil called an inductor and a capacitor. The inductance of the coil and the capacitance of the capacitor are adjusted to the resonance frequency to be transmitted.

The schematic circuit would look something like this.

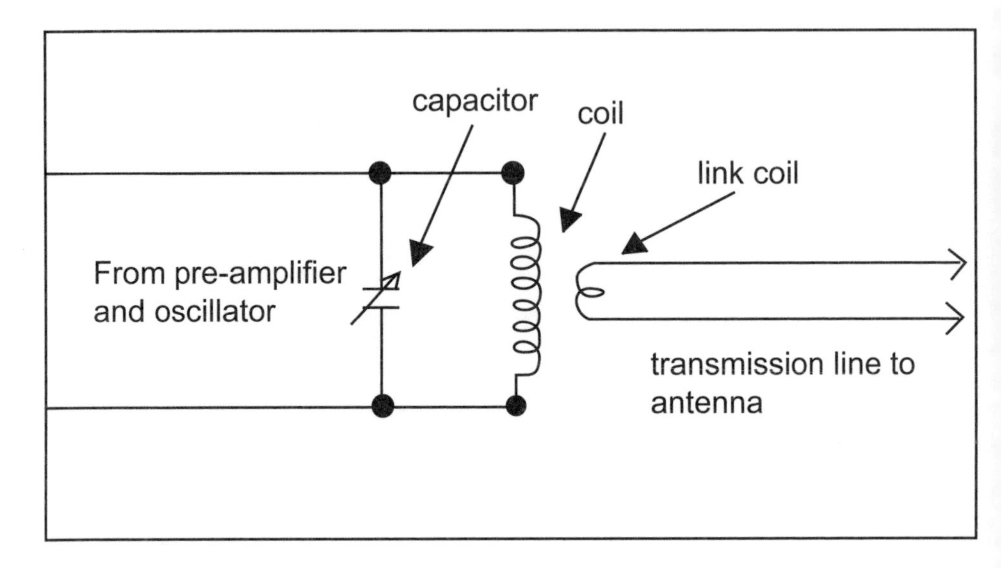

DRAWING 13-1

I am presenting this to you to let you know why I use the words “resonance acceptance”.

I need to explain something else before we can continue.

There is an instrument known as a “Grid-dip Oscillator”.

The Grid-dip Oscillator, GDO, is a small variable frequency transmitter. It also can be used as an ABSORPTION WAVEMETER. The Grid-dip Oscillator, GDO, is calibrated in frequency.

In the ABSORPTION MODE THE GRID-DIP OSCILLATOR, GDO, IS NOT TRANSMITTING A SIGNAL, but

will absorb energy at the frequency to which it is resonated.
For example: Suppose a transmitter is transmitting a signal at 7 MHz.
When the coil of the Grid-dip Oscillator, GDO, resonated to 7 MHz, is put near the coil of the transmitter, the Grid-dip Oscillator, GDO, will absorb energy from transmitter transmitting a signal at 7 MHz.

If the GDO is set to 6 MHz or 8 MHz it will not absorb the 7 MHz transmitted signal of the transmitter.

The GDO will absorb energy when its resonance frequency is the same as the resonance frequency of the transmitter.
See drawing 13-2.

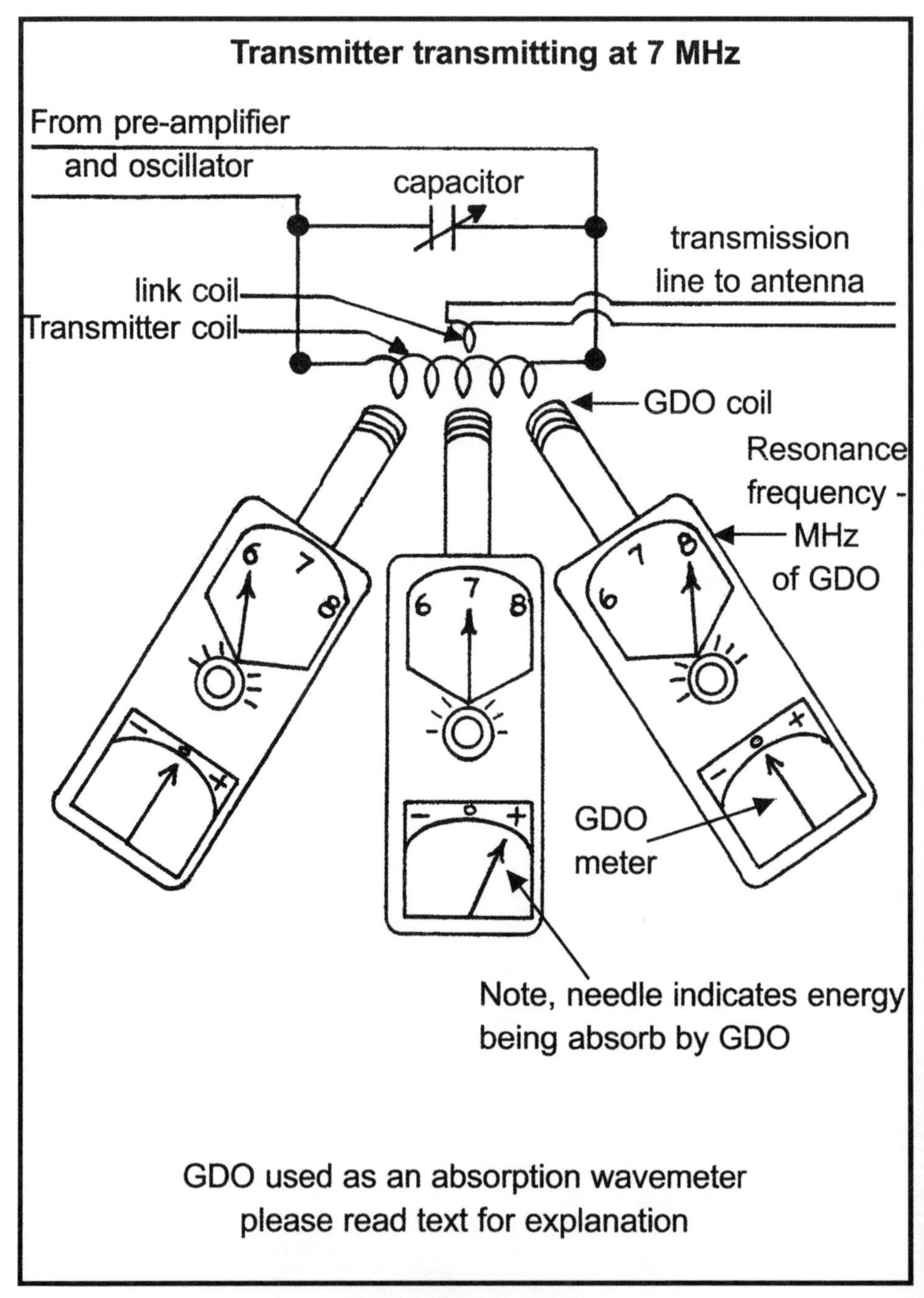
Transmitter transmitting at 7 MHz
From pre-amplifier and oscillator
capacitor
transmission line to antenna
link coil
Transmitter coil
GDO coil
Resonance frequency - MHz of GDO
6
7
8
–
o
+
GDO meter
Note, needle indicates energy being absorb by GDO
GDO used as an absorption wavemeter please read text for explanation

DRAWING 13-2

When the GDO is used as a GDO, that is it is TRANSMITTING A SIGNAL at a certain frequency – in this case a signal at 7 MHz – and the coil of the GDO is put near another coil in a non-energized circuit that is also resonance at 7 MHz – then the needle on the GDO meter will "dip", indicating that energy is being removed from the GDO by the external resonance circuit.

If the GDO is transmitting a signal at 6 MHz or 8 MHz, then the non-energized 7 MHz transmitter coil circuit will not absorb these signals.

See drawing 13-3 on next page

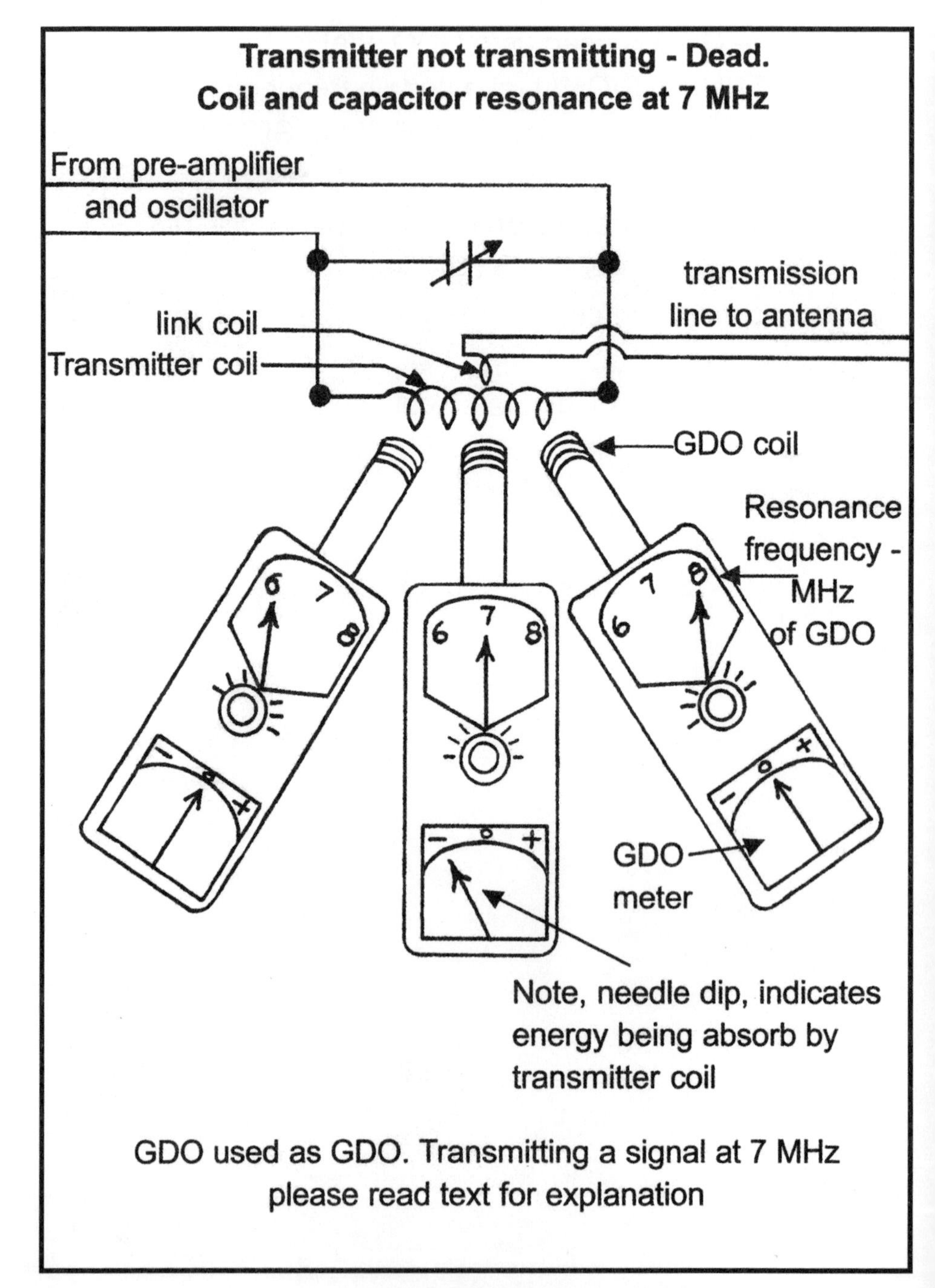
Transmitter not transmitting - Dead.
Coil and capacitor resonance at 7 MHz
From pre-amplifier
and oscillator
transmission
line to antenna
link coil
Transmitter coil
GDO coil
Resonance
frequency -
MHz
of GDO
6
7
8
GDO
meter
Note, needle dip, indicates
energy being absorb by
transmitter coil
GDO used as GDO. Transmitting a signal at 7 MHz
please read text for explanation

DRAWING 13-3

One more example of "resonance acceptance". Suppose a 10 watt transmitter in Hawaii is transmitting a signal at 7 MHz, and someone in Chicago, Illinois, U.S.A. wishes to receive this signal.

The person in Chicago sets his short-wave receiver to receive a 7 MHz signal. Then he connects an antenna wire to the receiver, say the wire is 150 ft in length (45.7 meters) and listens to the receiver and hears a very weak or no signal.

The signal being very weak, so the receiving antenna needs to be made resonant to the frequency of the transmitting signal, 7 MHz, in Hawaii.

There is a formula for determining what the receiving antenna length should be for 7 MHz. The length of the antenna should be 134 ft (40.8 meters). So the antenna wire is cut to the 134 ft. and now the 7 MHz signal from Hawaii is heard.

This again demonstrates "resonance acceptance".

NOTES

CHAPTER 14

– ELECTRICITY –
FLOW OF MAGNETISM? OR FLOW OF ELECTRONS?

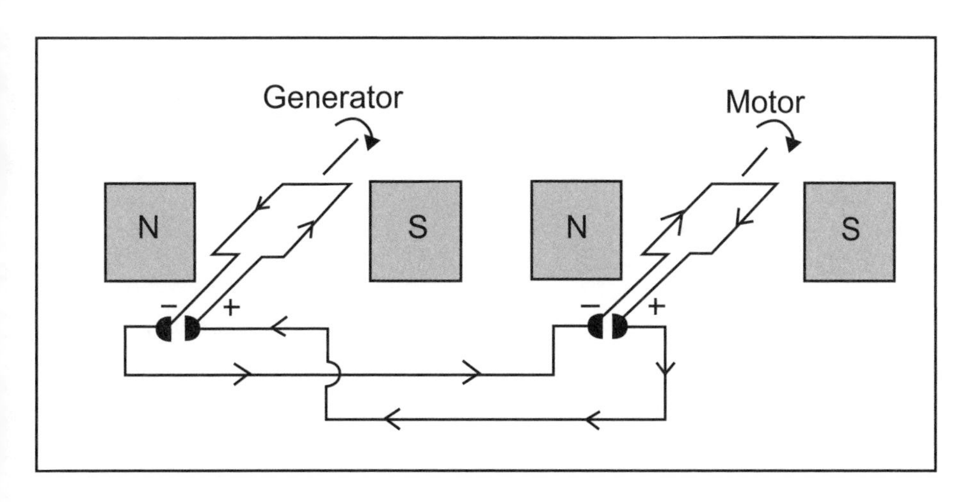

DRAWING 14-1

Assume we have a copper wire coil (armature) rotating in a magnetic energy field.

Rotation of the armature in the magnetic energy field causes the "Inter-Atomic-Magnetic-Energy" within the atoms to flow in the wire.

At the point of inception the "Inter-Atomic-Magnetic-Energy" level will be higher than the surrounding Inter-Atomic-Magnetic-Energy level, so the Inter-Atomic-Magnetic-Energy will move into this lower level.

As the rotation of the armature increases, more and more, it causes an increase of Inter-Atomic-Magnetic-Energy level.

This continuation of increase in Inter-Atomic-Magnetic-Energy level at point of inception causes the Inter-Atomic-Magnetic-Energy pressure to flow in the wire from high value to lower value.

At the far end of the wire the Inter-Atomic-Magnetic-Energy enters a coil in a magnetic field.

The Inter-Atomic-Magnetic-Energy causes a magnetic field to build up in the coil. This coil field will oppose or add to the existing pole field causing the coil to rotate.

Here we have presented a case for the flow of magnetism.

Now a case for electron flow.

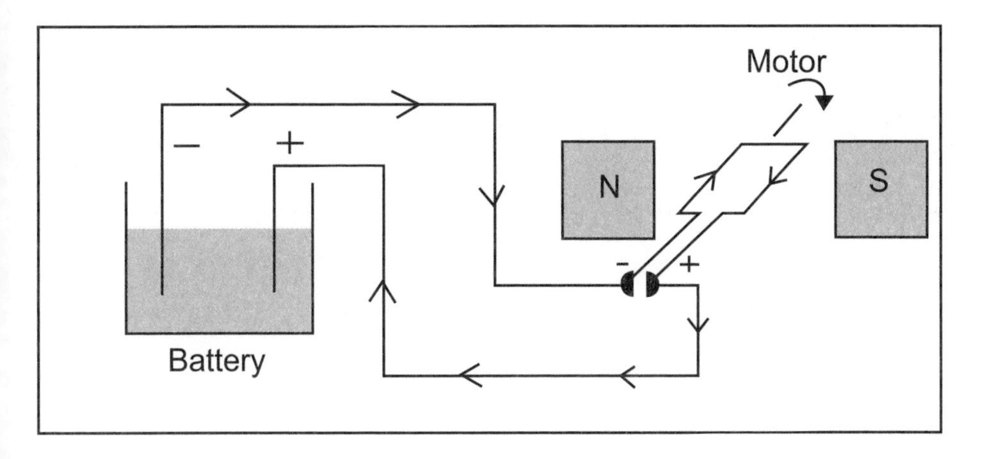

DRAWING 14-2

From the battery we have a flow of electrons produced by a chemical action. The electrons flow from battery to the armature coil of the motor. The flow of electrons in the coil create a magnetic field about the armature coil causing it to rotate in the motor's magnetic field.

So; here we have a flow of electrons doing the same as a flow of magnetism.

To me magnetism and electron movement are both forms of what we call electricity. Just different forms of the same, what we call electricity.

NOTES

CHAPTER 15

DOPPLER EFFECT

The reason for this chapter is that I want to present my interpretation of the doppler effect.

For example: using a moving 330 Hz/sec. signal pressure movement in air.

This is the frequency the stationary listener will hear when the 330 Hz/sec. is not moving.

Now, if the 330 Hz/sec. signal is moving toward the stationary listener – the stationary listener will hear the 330 Hz/sec. signal more times per second. Not at a higher frequency than the 330 Hz/sec. signal.

Ask yourself the question, why should the 330 Hz/sec. signal change its frequency just because it is moving?

If the signal is approaching the stationary listener – the 330 Hz/sec. signal will be heard more times per second by the stationary listener.

If the signal is receding from the stationary listener, the 330 Hz/sec. will be heard less times per second by the stationary listener.

The signal carrier moving forward or backward is not going to change the original frequency signal.

On the chart, I have put the stationary observer and the signals energies acumulation barrier at the end of the 4th second, although I could have stopped sooner or extended the chart.

Notice all signals movements energies accumulate at the same time at the end of the 4th second where I put the signals energies accumulation barrier.

As the moving time - distance, from the start, is extended, the signals energies accumulation barrier accumulates more and more energy.

If the amount of energy accumulated in the accumulation barrier is more than the medium that the energy is moving in can absorb, then the accumulated energies will be more than the medium can absorb and we have the sonic boom.

The stationary observer and the end of the 4th second

received four 330 Hz/sec. signals at the same time, not just one 330 Hz/sec. signal as it would be if the 330 Hz/sec. signal were not moving.

Notice that the 330 Hz/sec. signal energy starts to accumulate after “start” to end of the 4th second.

Actually the time I used, 4 seconds, may be broken down into any number of smaller time units.

During the entire process there is no change in frequency of the 330 Hz/sec. signal.

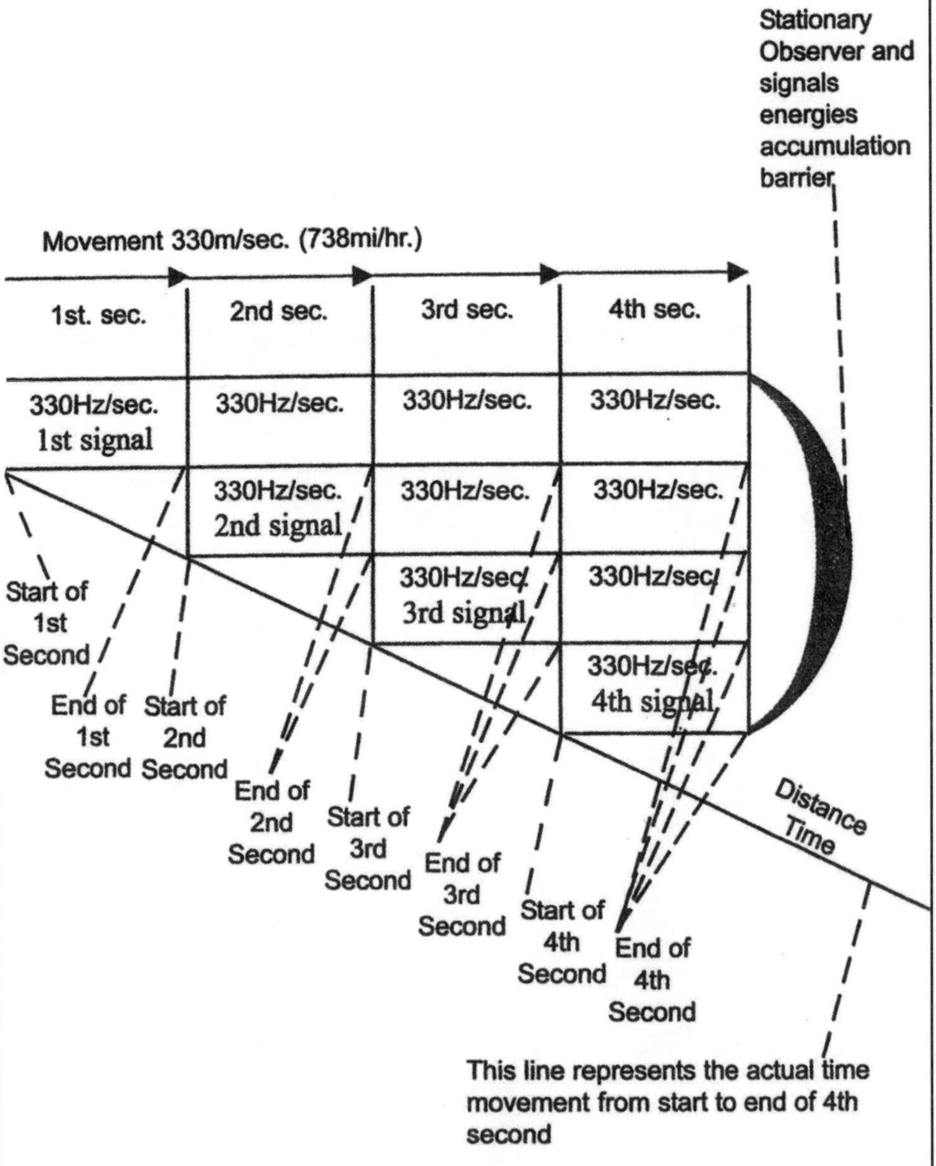

Stationary Observer and signals energies accumulation barrier

Movement 330m/sec. (738mi/hr.)

1st. sec.	2nd sec.	3rd sec.	4th sec.
330Hz/sec. 1st signal	330Hz/sec.	330Hz/sec.	330Hz/sec.
	330Hz/sec. 2nd signal	330Hz/sec.	330Hz/sec.
		330Hz/sec. 3rd signal	330Hz/sec.
			330Hz/sec. 4th signal

Start of 1st Second

End of 1st Second

Start of 2nd Second

End of 2nd Second

Start of 3rd Second

End of 3rd Second

Start of 4th Second

End of 4th Second

Distance Time

This line represents the actual time movement from start to end of 4th second

The signal energies accumulation barrier and stationary observer are receiving 4 - 330Hz/sec, Signals at the end of the 4th second.

DRAWING 15-1

CHAPTER 16

SPACE ENERGY AND OUR GALACTIC ENERGY AND EXTRA GALACTIC ENERGY

Now we need to study our galatic and extra galactic space energy and energy in general.

There are many forms of energy in our galactic and extra galactic systems.

Everything is in some form of energy. Motion is the movement of energy from one energy level to another energy level.

Many forms of our galactic and extra galactic space energy have been discovered in the last 100 years.

Some examples:

In 1931 amateur radio scientist W8JK, Karl Jansky, discovered radio signals coming from the Milky Way.
In 1937 amateur radio scientist W9GFZ, Grote Reber, made a

9 meter parabolic dish antenna for the study of galactic space energy. In 1941 he published the first radio energy map of the sky.

These two amateur radio scientists started radio astronomy.

More of our galactic and extra galactic energy discoveries are: solar wind, magnetic fields on the sun, our galactic and extra galactic magnetic field sources, ionized gas in the upper atmosphere, x-rays from outer space, cosmic rays from outer space, microwave cosmic radiation and others.

Our galactic and extra galactic energy sources have been discovered in the past 100 years. I believe more types of our galactic and extra galactic energy will be discovered in the future. So you can see that there is a lot of research to be done. We have just started.

Rotation of the Earth in the Glactic Space Medium Energy - GSME causes an electrical current to be produced in metallic objects here on Earth.

There is a possibility that this Glactic Space Medium Energy produces an electric current in the metal core of the Earth and form this action the Earth's magnetism is produced.

Now let us look into some known phenomena and work our way to my theory.

Two parallel wires carrying electric currents in the same direction attract each other.

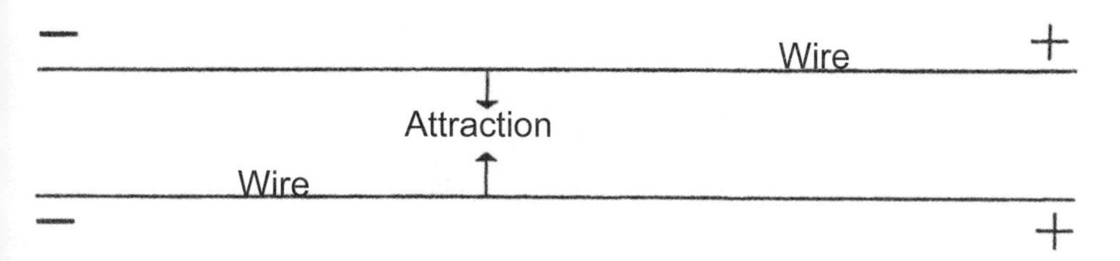

DRAWING 16-1

Two parallel wires carrying electric currents in the opposite directions repel each other.

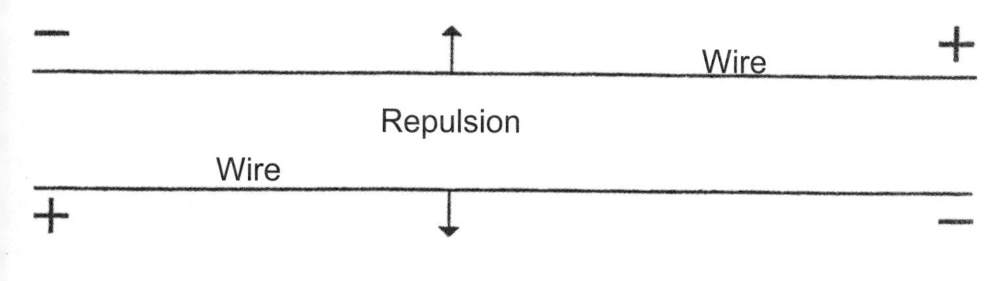

DRAWING 16-2

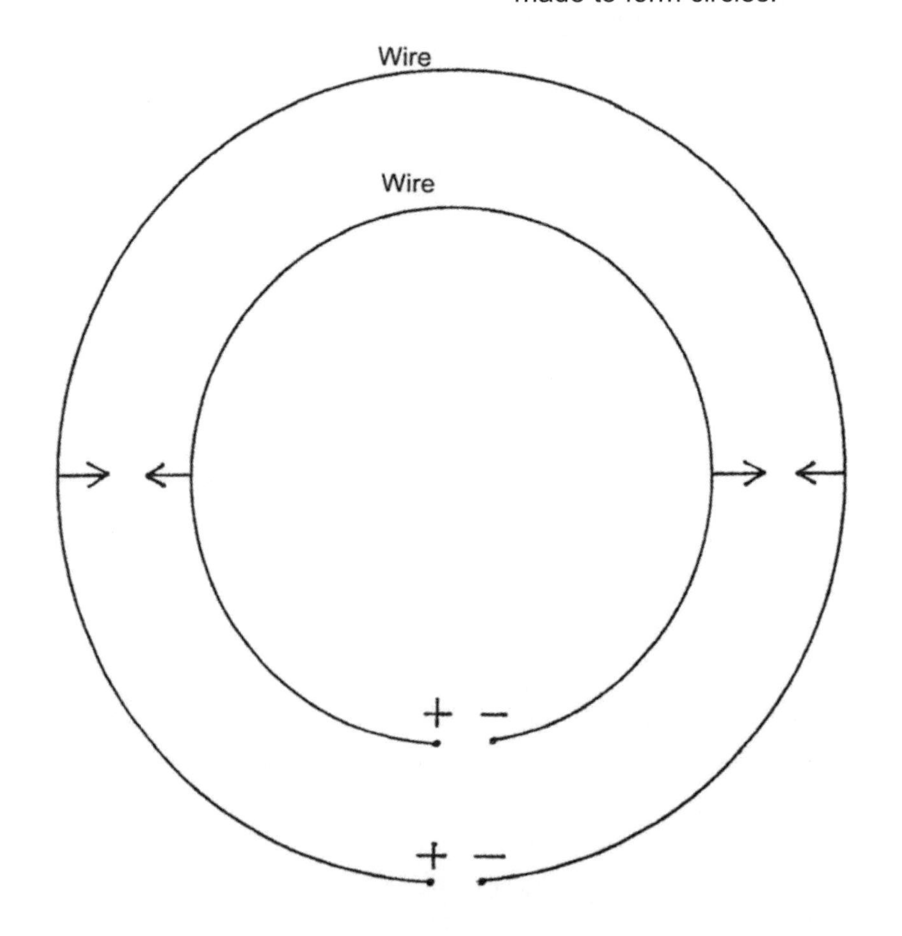
Here again parallel wires are made to form circles.
Wire
Wire
+
–
+
–
The arrows show the directions of attraction when electric currents are in the same direction.

DRAWING 16-3

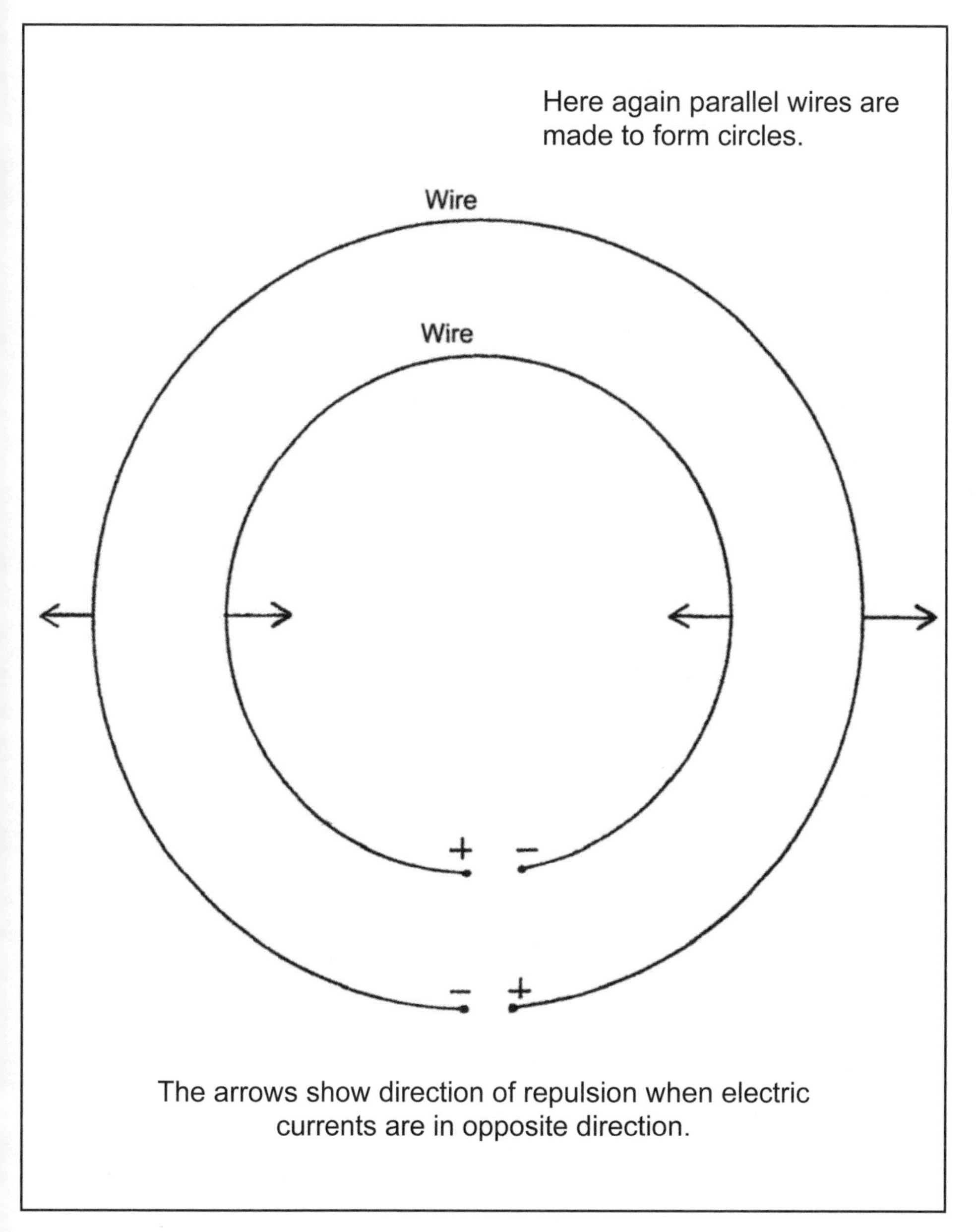
Here again parallel wires are made to form circles.
Wire
Wire
+
−
−
+
The arrows show direction of repulsion when electric currents are in opposite direction.

DRAWING 16-4

Now lets condense the wires so each wire becomes one lump of material. Each lumb of material goes in its own elliptical orbit. These lumps of material we will call planets. P1 being Earth and P2 being Mars.

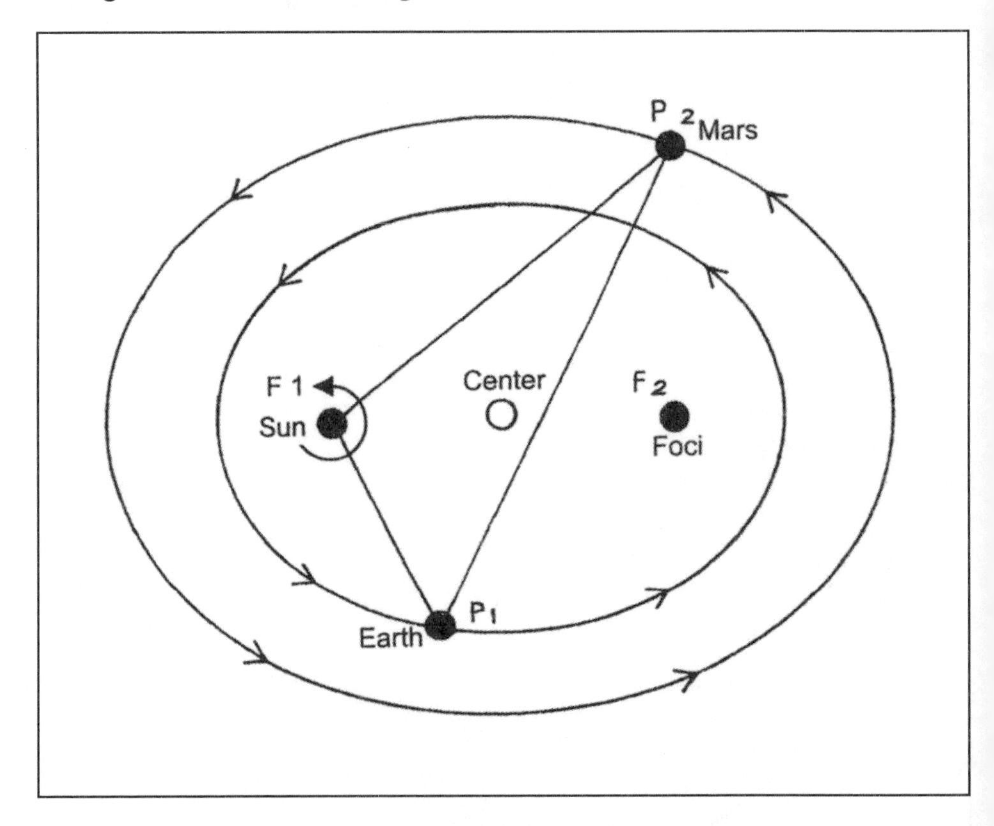

DRAWING 16-5

These two planets will do to illustrate how the planets and sun attract. Notice all are moving in a counter clockwise direction.

SPACE ENERGY AND OUR GALACTIC ENERGY AND EXTRA GALACTIC ENERGY

The planets move through our galactic space energy while going in their elliptical orbits about the sun. While this is taking place, our solar system is also moving clockwise in our galaxy.

My thinking is that this Galactic Space Medium Energy our sun and planets are moving in is part of our galactic system.

The movements that we are concerned with are:

1. Counter clockwise movement of our sun and planets in our solar system.
2. Clockwise movement of our solar system in our galaxy.

NOTE: In Chapter 17 I explain the amount of attraction energy value of the planets and the sun.

NOTE: In Chapter 18 I explain the repulsion energy value of the sun.

NOTE: In Chapter 19 I explain “Our Sun Repulsion Energy and Attraction Energy”.

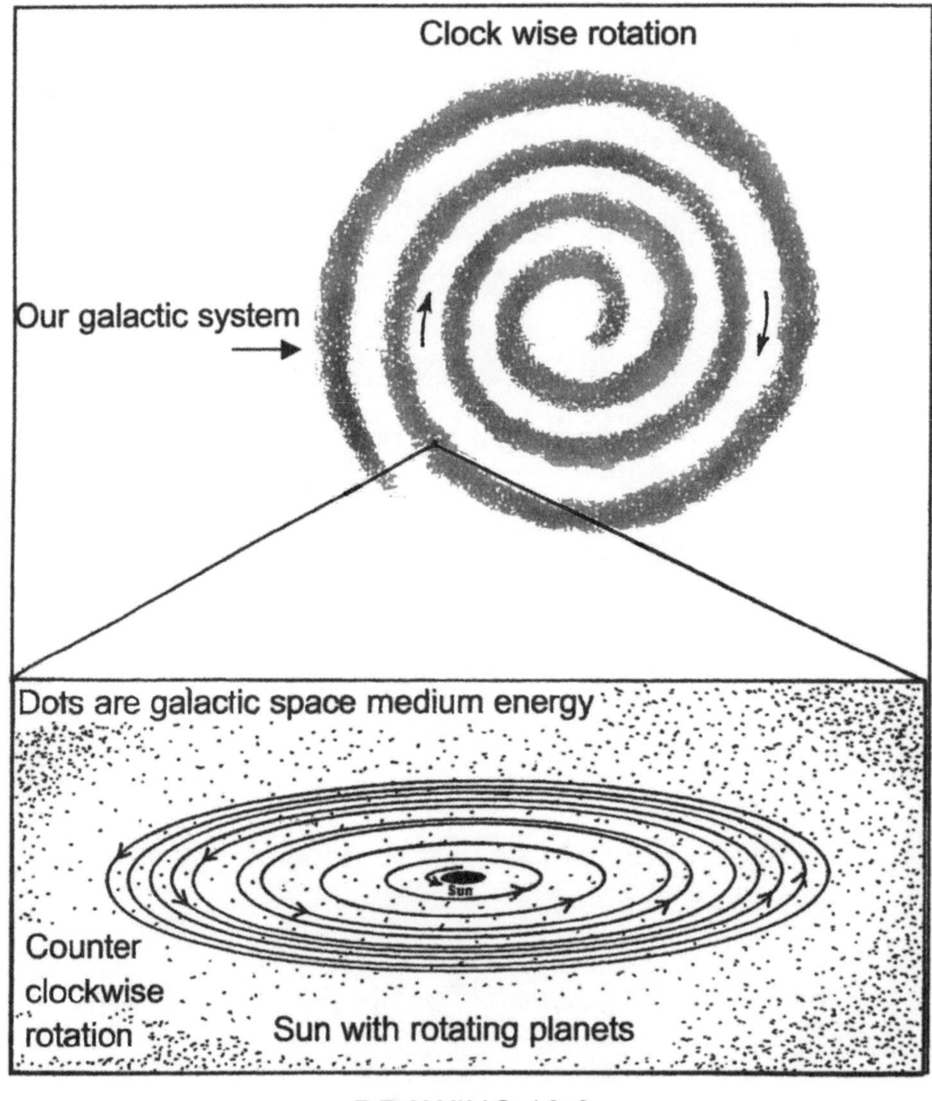

DRAWING 16-6

Our solar system "shown in the blocked area," moves clockwise in our galactic system.

CHAPTER 17

ATTRACTION ENERGY VALUE OF THE NINE PLANETS AND THE SUN

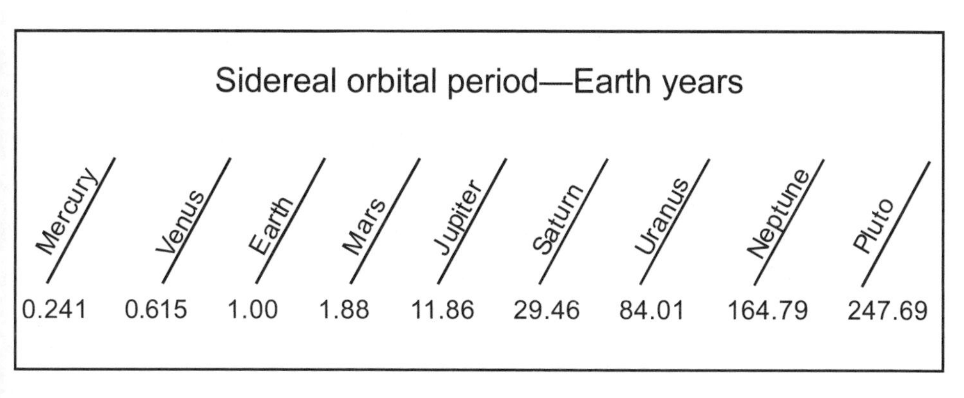

DRAWING 17-1

$\frac{247.69 \text{ Pluto}}{0.241 \text{ Mercury}}$ = 1,027.76 number of times Mercury will orbit the sun during Pluto's orbit.

$\frac{247.69 \text{ Pluto}}{0.615 \text{ Venus}}$ = 402.75 number of times Venus will orbit the sun during Pluto's orbit.

$\frac{247.69 \text{ Pluto}}{1.00 \text{ Earth}}$ = 247.69 number of times Earth will orbit the sun during Pluto's orbit

$\frac{\text{247.69 Pluto}}{\text{1.88 Mars}}$ = 131.75 number of times Mars will orbit the sun during Pluto's orbit

$\frac{\text{247.69 Pluto}}{\text{11.86 Jupiter}}$ = 20.88 number of times Jupiter will orbit the sun during Pluto's orbit

$\frac{\text{247.69 Pluto}}{\text{29.46 Saturn}}$ = 8.41 number of times Saturn will orbit the sun during Pluto's orbit

$\frac{\text{247.69 Pluto}}{\text{84.01 Uranus}}$ = 2.95 number of times Uranus will orbit the sun during Pluto's orbit

$\frac{\text{247.69 Pluto}}{\text{164.79 Neptune}}$ = 1.50 number of times Neptune will orbit the sun during Pluto's orbit

$\frac{\text{247.69 Pluto}}{\text{247.69 Pluto}}$ = 1.00 number of times Pluto will orbit the sun during Pluto's orbit

The above chart shows the number of times each planet orbits the sun compared to the orbital period of Pluto.

Pluto having the longest orbital time distance in years. We need to find the attractive energy value for each planet, so we can get the total attractive energy valve for our solar system. For example: Mercury will make 1027.76 orbits about the sun while Pluto makes 1 orbit about the sun.

ATTRACTION ENERGY VALUE OF THE NINE PLANETS AND THE SUN

Now we need to take the Mass of each planet and multiply this by the number of times each planet orbits the sun compared to the orbital period of Pluto.

Mass X Number of times planet will orbit the sun during Pluto's orbit = Attractive energy value kg/247.69 earth years

Mercury
3,300 x 10^{20} kg x 1,027.76 = 3,391,608 x 10^{20} kg/247.69 years

Venus
48,700 x 10^{20} kg x 402.75 = 19,613,925 x 10^{20} kg/247.69 years

Earth
59,700 x 10^{20} kg x 247.69 = 14,787,093 x 10^{20} kg/247.69 years

Mars
6,420 x 10^{20} kg x 131.75 = 845,835 x 10^{20} kg/247.69 years

Jupiter
18,990,000 x 10^{20} kg x 20.88 = 396,511,200 x 10^{20} kg/247.69 years

Saturn
5,680,000 x 10^{20} kg x 8.41 = 47,768,800 x 10^{20} kg/247.69 years

Uranus
868,000 x 10^{20} kg x 2.95 = 2,560.6 x 10^{20} kg/247.69 years

Neptune
1,020,000 x 10^{20} kg x 1.50 = 1,530,000 x 10^{20} kg/247.69 years

Pluto
125 x 10^{20} kg x 1 = 125 x 10^{20} kg/247.69 years

	Attractive Energy Value
Mercury	3,391,608 x 10^{20} kg
Venus	19,613,925 x 10^{20} kg
Earth	14,787,093 x 10^{20} kg
Mars	845,835 x 10^{20} kg
Jupiter	396,511,200 x 10^{20} kg
Saturn	47,768,800 x 10^{20} kg
Uranus	2,561 x 10^{20} kg
Neptune	1,530,000 x 10^{20} kg
Pluto	125 x 10^{20} kg

Total 484,451,022 x 10^{20} kg/247.69 earth years

Or

0.0484451022 x 10^{30} kg/247.69 earth years

This is the ATTRACTIVE VALUE FOR THE NINE PLANETS. Not much.

ATTRACTIVE VALUE FOR THE SUN IS 4,973 x 10^{30} kg/247.69 earth years

ATTRACTION ENERGY VALUE OF THE NINE PLANETS AND THE SUN

Attractive Energy Value For Our Sun

Sun Orbital Period in Years

The sun roates once on its axis every 36 earth days, counterclockwise.

$$\frac{\text{36 earth days—sun rotation on its axis}}{\text{365 earth days—earth rotation years on its axis for 1 year.}} = 0.0986 \text{ years, sun rotation on its axis in earth years.}$$

Pluto Orbital Period

$$\frac{\text{247.69 Pluto}}{\text{0.0986 Sun}} = 2{,}512.06$$

0.0986 Sun: sun rotation on its axis in earth years

2,512.06: Number of times the sun will rotate on its axis during Pluto's orbit

Mass of sun **x** number of times the sun will rotate on its axis during Pluto's orbit

1.98×10^{30} kg **x** 2,512.06 = 4,973.878 **x** 10^{30} kg
Counter Clock Wise Attractive Energy Value of our sun.

The attractive value for the sun and the attractive values of the planets take place during the time distance of 247.69 earth years (Pluto's orbit distance in earth years)

These are all Attractive Values as they are COUNTER CLOCKWISE ROTATIONS.

With respect to our galaxy's clockwise rotation.

NOTES

NOTES

CHAPTER 18

REPULSION ENERGY VALUE OF THE SUN

There must be many types of galactic energy in our galaxy space.

When a Mass moves through this galactic space energy an energy is produced in the Mass that is an attraction or repulsion energy.

All Mass moving in the same direction in the same galactic energy will have an attractive energy produced in them and will attract each other. This is gravity.

If two Masses are moving in the same galactic energy in opposite directions, the Masses will have opposing energies and will repel each other.

The sun rotates on its axis once every 36 earth days, counter clockwise.

Sun orbital velocity in our galaxy is 220 km/second

220 km/second x 60 = 13,200 km/minute
13,200 km/minute x 60 = 792,000 km/hour
792,000 km/hour x 24 = 19,008,000 km/day
19,008,000 km/day x 365 = 6,937,920,000 km/year

Sun orbital velocity in our galaxy is 6,937,920,000 km/earth year.

19,008,000 km/day Sun orbital velocity **x** 365 earth days/year = 6,937,920 **x** 10^3 km/year sun orbital velocity in our galaxy

6,937,920 **x** 10^3 km/year (sun orbital distance for 1 year) **x** 247.69 years (Pluto's orbit in years.)
= 1,718,453,405 **x** 10^3 km/247.69 years
sun orbit distance for 247.69 years

1,718,453,405 **x** 10^3 km/247.69 years
x 1.98 **x** 10^{30} kg sun Mass
= 3,402,537,742,000 **x** 10^{30} kg/km/247.69 years
sun repulsion energy
or
3,402,537,742 **x** 10^3 **x** 10^{30} kg/km/247.69 years

3,402,537,742,000 $\times 10^{30}$ kg/km/247.69 years
Sun Repulsion Energy

-4,973 $\times 10^{30}$ kg/km/247.69 years
Sun Attraction Energy

3,402,537,737,027 $\times 10^{30}$ kg/km 247.69 years Difference in Favor Sun Repulsion Energy

All this could be done other ways or by some division could be put on a one earth year basis.

So, I will put the sun repulsion, sun attraction and difference on a one earth year basis.

$$\frac{3{,}402{,}537{,}742{,}000 \times 10^{30} \text{ kg/km}}{247.69 \text{ years}} = 13{,}737{,}081{,}600 \times 10^{30} \text{ kg/km}$$

sun repulsion energy for 1 earth year

$$\frac{4{,}973 \times 10^{30} \text{ kg/km}}{247.69 \text{ years}} = 20.08 \times 10^{30} \text{ kg/km}$$

sun attraction energy for 1 earth year

13,737,081,600.00 $\times 10^{30}$ sun repulsion energy
-20.08 $\times 10^{30}$ sun attraction energy
13,737,081.579.92 $\times 10^{30}$ difference, repulsion in favor for sun

All 3 figures above are for 1 earth year.

NOTES

CHAPTER 19

OUR SUN, REPULSION ENERGY AND ATTRACTION ENERGY

Now What! The numbers do not support modern gravity theory.

Now all of a sudden our sun has much more repulsion energy than attraction energy.

I and everyone else thought the sun was the big attractor, meaning that the sun, would eventually attract its planets and other surrounding objects into itself.

After doing this research work and coming up with these figures, I have changed my thinking on the subject.

Now, the sun with its overwhelming repulsion energy must be pushing the planets and surrounding objects away from itself, not attracting as we all thought.

According to modern day gravity theory, the sun has a lot of

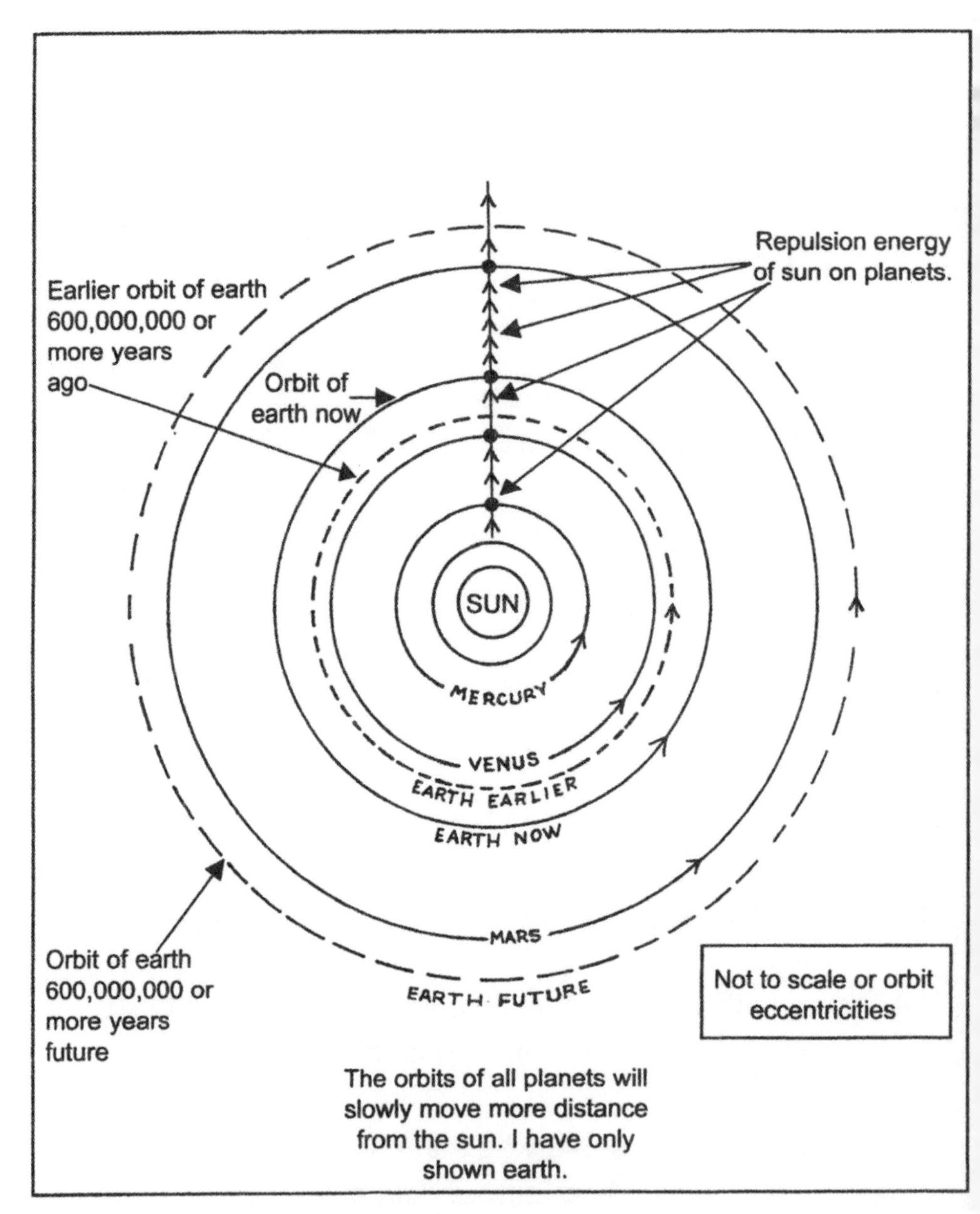
Repulsion energy of sun on planets.
Earlier orbit of earth 600,000,000 or more years ago
Orbit of earth now
SUN
MERCURY
VENUS
EARTH EARLIER
EARTH NOW
MARS
EARTH FUTURE
Orbit of earth 600,000,000 or more years future
Not to scale or orbit eccentricities
The orbits of all planets will slowly move more distance from the sun. I have only shown earth.

DRAWING 19-1

attraction for the planets and the planets also have attraction for the sun.

According to my theory on Repulsion Energy from the sun, places the planets much closer to the sun than they are now.

The Repulsion Energy of the sun has moved the orbits of the planets further and further from the sun as distance-time move on, or really as our solar system moves in our galaxy.

Please refer to drawing 19-1 for more detail information.

The earth, and other planets, must have been closer to the sun billions of years ago. At that time, they must have had a faster orbital period and a faster axis rotation, even the axis of inclination (23-1/2 degrees from vertical to plane of earth's orbit) may have been different.

Plenty of sunshine, heat, rain, lighting (to produce nitrogen fertilizer). All vegetative matter grew very rampant, larger animals were in abundance as were many other forms of life.

Some Dinosaurs were 65 feet (19.8 meters) to 85 feet (25.9 meters) long and weighing as much as 10 short tons (9.07 metric tons) or more.

Just think of how much food larger animals of this size require each day.

This was earth's "Golden Era".

As the repulsion energy from the sun caused the orbits to slowly move to larger orbits, all this "Golden Era" for the earth is in the past. Could this have been the cause of the Dinosaur extinction?

If one compares the earth of today with the earth of its "Golden Era," one can see that its future does not look too good.

In the future, billions of years from now, the oceans will be frozen and all life will be extinct. The earth will just be another lump of matter in a more distant orbit from the sun.

From earth fossils, we have a record to show that the earth today is almost a desert compared to the earth of the "GOLDEN ERA."

INDEX

Chapter 1- 4

I

K

L

M

N

P

Q

R

S

T

V

W

Y

INDEX

Chapter 5 -19

A

B

C

D

E

S

T

V

W